Axel Stockhausen

Optimization of Hot-Electron Bolometers for THz Radiation

Optimization of Hot-Electron Bolometers for THz Radiation

BAND 008

Karlsruher Schriftenreihe zur Supraleitung

Herausgeber

Prof. Dr.-Ing. M. Noe
Prof. Dr. rer. nat. M. Siegel

Eine Übersicht über alle bisher in dieser Schriftenreihe
erschienene Bände finden Sie am Ende des Buchs.

Optimization of Hot-Electron Bolometers for THz Radiation

by
Axel Stockhausen

Dissertation, Karlsruher Institut für Technologie (KIT)
Fakultät für Elektrotechnik und Informationstechnik, 2013
Hauptreferent: Prof. Dr. Michael Siegel
Korreferent: Prof. Dr. Paul Seidel

Impressum

Karlsruher Institut für Technologie (KIT)
KIT Scientific Publishing
Straße am Forum 2
D-76131 Karlsruhe
www.ksp.kit.edu

KIT – Universität des Landes Baden-Württemberg und
nationales Forschungszentrum in der Helmholtz-Gemeinschaft

KIT Scientific Publishing 2013
Print on Demand

ISSN 1869-1765
ISBN 978-3-7315-0066-7

Optimization of Hot-Electron Bolometers
for THz Radiation

Zur Erlangung des akademischen Grades eines
DOKTOR-INGENIEURS
an der Fakultät für
Elektrotechnik und Informationstechnik
der Universität Karlsruhe (TH)
genehmigte
DISSERTATION

von
Diplom-Physiker Axel Stockhausen
geboren in Heilbronn

Tag der mündlichen Prüfung: 24. Juni 2013

Hauptreferent: Prof. Dr. Michael Siegel

Korreferent: Prof. Dr. Paul Seidel

Preface

This dissertation is the result of my work at the *Institut für Mikro- und Nanoelektronische Systeme* which was conducted at the *Karlsruhe Institute for Technology* (KIT). This thesis would not have been possible without the help of many different people whom I want to thank in the following.

Danksagung

Diese Arbeit wäre nicht ohne die Unterstützung von vielen Menschen möglich gewesen, denen ich auf diesem Wege noch einmal meinen Dank aus-sprechen möchte. Zu Beginn möchte ich mich bei Herrn Professor Siegel dafür bedanken, dass er es mir ermöglicht hat, diese Arbeit durchzuführen, bei Herrn Professor Seidel für die Übernahme des Koreferats und bei Dr. Konstantin Ilin für die Betreuung während der Promotion.

Ein besonderes Dankeschön gilt Herrn Doktor Wünsch für die beständige seelische und moralische Unterstützung, die mir sehr geholfen hat, diese Arbeit zu vollenden. Besonders danke ich auch Dr. Alexander Scheuring möchte für die langjährige und vor allem sehr unterhaltsame Zusammenarbeit. Des Weiteren bedanke ich mich bei allen Mitarbeitern und Studenten, mit denen ich gemeinsam am Institut gearbeitet habe. Namentlich erwähnen möchte ich Dagmar Henrich, Matthias Hofherr, Max Meckbach, Petra Thoma, Gerd Hammer, Matthias Arndt, Christoph Kaiser, Hansjürgen Wermund, Karlheinz Gutbrod, Frank Ruhnau, Alexander Stassen, Erich Crocoll und Doris Duffner.

Nicht zuletzt geht ein spezieller Dank an meine Eltern, Mona und Elmar Stockhausen, an meine Brüder Philipp und Felix sowie an meine Freundin Carolin Henrich, die mich über die lange Zeit der Doktorarbeit immer unterstützt haben, und ohne die ich wahrscheinlich nie fertig geworden wäre.

Zusammenfassung

Diese Arbeit beschäftigt sich mit der Optimierung von Hot-Electron Bolometern für den THz-Frequenzbereich. Hot-Electron Bolometer sind hochsensible, rauscharme Strahlungsdetektoren die in vielen Bereichen der Forschung Anwendung finden. Gerade im Bereich der Astrophysik sind besonders rauscharme Detektoren mit einer hohen Bandbreite von mehreren Gigahertz stark nachgefragt. Hot- Electron Bolometer machen sich den namensgebenden "Hot- Electron" Effekt zu nutze. In Systemen in denen der Hot- Electron Effekt vorkommt, kann das Elektronensystem des Detektormaterials sich vom Phononensystem entkoppeln. Beide Systeme können dann in einem eigenen thermischen Gleichgewicht sein. Durch Ausnutzung dieses speziellen Zustands können Detektoren hergestellt werden, die über eine hohe Bandbreite und Empfindlichkeit verfügen. Die so hergestellten Hot-Electron Bolometer bestehen aus einer ultradünnen Niobnitrid Schicht, die auf ein Silizium- oder Saphirsubstrat mittels reaktivem Magnetronsputtern aufgebracht wird. Diese Schicht wird mit Hilfe Nanolithographie zu einem Detektorelement strukturiert. Zur Einkopplung der THz-Strahlung wird eine planare Goldantenne, die direkt auf der Detektorschicht abgeschieden wird, benötigt. Diese wird mittels Lift-Off strukturiert.

Ziel dieser Arbeit ist es den Fertigungsprozess so zu optimieren und zu verstehen, dass die Herstellung von Detektoren mit einer hohen Bandbreite und einem geringen Rauschen realisiert werden kann. Es wurden unterschiedliche Schritte durchgeführt, um dieses Ziel zu erreichen. In einem ersten Schritt wurde die Herstellung der einzelnen supraleitenden Schichten optimiert, und die supraleitenden Eigenschaften der abgeschiedenen Dünnfilme untersucht. Um den Einfluss der unterschiedlichen Schichten aufeinander zu bestimmen, wurden Mehrfachschichtsysteme auf ihre Eignung für die Herstellung von rauscharmen Detektoren untersucht. Hier stellte sich heraus, dass die normalleitende Schicht der Goldantenne die Supraleitung im Detektorelement unterdrückt. Um diesen Effekt zu verhindern, wurden supraleitende Pufferschichten zwischen der Detektorschicht und der Antennenschicht entwickelt. Die auf diese Weise hergestellten Detektoren wiesen keine Unterdrückung der Supraleitung in ihrem Detektorelement auf. Bei der Charakterisierung der Rauschtemperatur dieser Detektoren wurden Rekordwerte erzielt. Dies zeigt, dass die durchgeführte Optimierung zu einer verbesserten Leistungsfähigkeit der Detektoren geführt hat.

In einem zweiten Schritt wurde die thermische Ankopplung des supraleitenden Dünnfilms an das verwendete Siliziumsubstrat untersucht. Die thermische Kopplung zwischen Substrat und supraleitendem Dünnfilm ist ein Kernparameter, um die Empfindlichkeit und Antwortzeit von Bolometern einzustellen. Um diese thermische Kopplung zu untersuchen, wurde der supraleitende Dünnfilm auf einen freistehenden Siliziumsockel mit variabler Höhe aufgebracht. Hier konnte eine starke Abhängigkeit der thermischen Kopplung von der Höhe des Sockels festgestellt werden. Um diesen Effekt zu erklären, wurde ein Modell entwickelt, das die gewonnenen Ergebnisse mit einer sehr guten Übereinstimmung beschreiben kann. Aufgrund dieser Ergebnisse wurden Detektoren hergestellt deren Detektorelement auf einem Siliziumsockel liegt. Bei der Charakterisierung der Antwortzeiten der Detektoren konnte eine starke Verlängerung der Antwortzeit festgestellt werden.

Abschließend wurde die Funktion der im Rahmen dieser Arbeit hergestellten Detektoren in Experimenten überprüft. In einem ersten Experiment wurde der Detektor zur Messung von kohärenter Synchrotronstrahlung im Terrahertz-Bereich verwendet. Die vom Synchrotron aus gesendeten Pulszüge konnten von dem Bolometer mit hoher Genauigkeit gemessen werden. Aus der Form der Einzelpulse ließ sich weiterhin ein Rückschluss auf die Antwortzeit des Detektors machen. In einem zweiten Experiment wurde der Detektor zur Charakterisierung eines Terrahertz-Quantenkaskadenlasers verwendet. Normalerweise werden für solche Experimente Golay-Zellen oder langsame SiGe-Bolometer verwendet. Durch die Verwendung eines schnellen Hot-Elektron Bolometers war es möglich nicht nur die mittlere Leistung des Quantenkaskadenlasers in einem Puls zu bestimmen. Zum ersten Mal war es in diesem Experiment möglich den zeitlichen Verlauf der Emission des Quantenkaskadenlasers zu messen, was für unsere Kooperationspartner zu weiteren Erkenntnissen für die Entwicklung dieser Laser geführt hat.

Zusammengefasst kann gesagt werden, dass die im Rahmen dieser Arbeit entwickelten

Detektoren, sehr gute Leistungsdaten in Laborexperimenten gezeigt haben. Weiterhin konnten sie bereits erfolgreich für über die klassischen Anwendungsgebiete hinausgehenden Experimente eingesetzt werden. Für die Zukunft sind viele weitere Anwendungen für diese Detektoren in den unterschiedlichsten Feldern denkbar.

Contents

Contents

1 Introduction

In recent years research in the field of terahertz radiation detection has seen a lot of attention. Astrophysics, Spectroscopy, medicine and security applications are only a few areas in science and technology that show an increased interest in measurements in this frequency range.

The THz frequency range from 0.1 THz to roughly 10 THz is of particular interest for these fields of research. In this frequency range lie a lot of vibration and absorption lines of molecules which makes it interesting for deep space radio astronomy and for spectroscopic analysis. It is also the frequency range where the absorption in the earths atmosphere is very strong. This calls for very sensitive detectors and high altitude or airborne experiments. Also THz radiation is damped only weakly by interstellar dust which further increases the usability for astronomy applications.

THz radiation can also pass through many materials used for packaging and clothing which, in recent years, attracted the interest of security agencies all around the world. By using sensitive THz detectors it is possible to measure the THz emission from a human body. Concealed objects will (in contrast to the clothing which is transparent) show a different THz emission than the rest of the body and thus can be detected even if they are beneath clothes. It is also interesting for non-destructive material testing where transmission measurements of THz radiation can reveal internal inclusions inside materials.

To be able to observe any of these effects it is important to have very sensitive detectors which can operate in this frequency range [1]. Below 1 THz the superconductor-isolator-superconductor (SIS) mixer shows the lowest noise equivalent power and are accepted as the most sensitive detectors in this the frequency range with noise down to the quantum limit. The cut-off frequency of the SIS mixer depends on the energy gap of the superconducting material used. Thus they can only be operated up to a certain gap frequency, for example 700 GHz for Niobium devices [2], which are the most common SIS mixers. At this frequency they show a noise temperature of $T \approx 100$ K. Above the gap frequency the noise temperature of these devices increases drastically. Using a different material like niobium nitride with a higher energy gap the expected theoretical limit is about 1.4 THz [3]. This limits the application of SIS devices in the THz frequency range.Due to the non-availability of fast detectors and radiation sources in the region between 1 THz and 10 THz, this frequency range has often been described as the "THz-gap" in the literature. Due to development of new kinds of detectors in the last decade the THz-gap is no longer existent [4]. To achieve ultimate sensitivity and fast signal response or mixer operation superconducting detectors are required. For frequencies above 1 THz superconducting detectors which are do not depend on the superconducting energy gap are required. One possible kind of detector for this application is the superconducting hot-electron bolometer. It is a fast superconducting bolometric detector that can be operated as a direct radiation detector and also as a mixer depending on the requirements of the application. It has been shown by several groups that HEB detectors can be used to detect radiation up to several THz [5]. HEB detectors are the most sensitive detectors in this frequency range and can reach noise temperatures of close to ten times the quantum limit (see e.g. figure 4 in [6]). The goal nowadays is to develop detectors with the smallest possible noise temperature. Such hot-electron bolometer devices might achieve ultimate sensitivity close to the quantum noise level.

Structure of the work

In chapter 2 of this work the basic operation principle of a bolometer is explained and the special case of the hot-electron bolometer is discussed and subsequently noise sources in bolometers are described. In a further part the different detection principles of direct and heterodyne detection are introduced and the need for an antenna coupling for the THz HEB is explained. The last part of chapter 2 focuses on the requirements and challenges for the development of HEB detectors.
The third chapter the focus is on the development and optimization of thin film deposition for the detector film and the antenna structure. The very important aspect of interfaces between the different films that make up the bolometer, and the interplay

between these films is discussed in the light of the proximity effect theory. The chapter closes with the characterization of a HEB mixer fabricated using optimized thin films and interfaces.

In the fourth chapter the modification of the thermal coupling between the thin super-conducting detector film and the thermal bath is discussed. A method to decouple the thin film from the thermal bath is developed. By putting the detector on a silicon mesa structure the phonon escape path gets restricted. A model is developed to describe the experimental measurements of thin films on different aspect ratios of silicon mesas. The chapter closes with the characterization of a detector fabricated sitting on a high silicon mesa structure.

The fifth chapter of this work focuses on the application of detectors fabricated in this work in large scale experiments. Therefore the sensitivity and the noise equivalent power of the detectors was first characterized using an in-house measurement system. In the second part the characterization of coherent synchrotron radiation with these detectors is demonstrated and from the data the speed of the detector is evaluated. In the last part of this chapter the detectors are used to characterize the time dependent power output of a quantum cascade laser.

The work closes in chapter 6 with a summary about what was achieved in this work and a short outlook.

2 Bolometric detectors for the THz spectrum

Bolometers are radiation power detectors. They measure the power of the radiation through absorption and a thus resulting change of a measurable quantity. The most commonly used bolometers are resistive bolometers. They use a change in resistance induced by absorption of radiation and measure the resulting current/voltage change. This change is proportional to the absorbed power.

Most detectors used in the optical and far infrared range as well as detectors developed for the microwave frequency range show only weak performance in the THz frequency range. The radiation detection of bolometric detectors in contrast is not frequency restricted as long as there is a way to absorb the incoming radiation.This marks bolometers as interesting candidates for radiation detection in the THz frequency range The bolometer principle is already quite old, being developed at the end of the 19th century. The first bolometer was developed by S. P. Langley in 1878 [7] as a tool to measure the infrared spectrum of the sun. He used a setup where two identical thin platinum stripes covered with carbon black lay close to each other. Only one of them was irradiated by light and warmed up. The change in resistance was measured with a Wheatstone bridge [8]. By using different absorber materials it was, since then, possible to fabricate bolometers for various radiation frequencies.

In this chapter we will first describe the basic operation principle of a bolometer. The most important figures of merit like sensitivity and speed of the bolometer are derived and explained. The special case of the hot-electron bolometer and the superconducting hot-electron bolometer is introduced and explained. In the second part we will quickly introduce two different modes of operation for radiation detectors that can be used with hot-electron bolometers, namely the direct detection and the heterodyne detection. In a subsequent section the need for coupling of the superconducting bolometer to an antenna will be discussed. In the last part of the chapter the requirements for a superconducting hot electron bolometer for the THz regime will be given.

2.1 Basics of bolometers operation

Bolometers are radiation detectors which have been widely used as sensitive detectors for thermal radiation since their development in the lat 19$^\text{th}$ century. The basic setup of the classical bolometer didn't change since then. A schematic view of the basic components of a bolometer is shown in figure 2.1.

The bolometer consists of an absorber material that converts the incoming radiation P_inc into heat. This absorber is coupled to the body of the bolometer. When radiation is absorbed by the absorber the produced heat is transferred to the bolometer body. The bolometer body heats up depending on the amount of power absorbed by its heat capacity C. The temperature change of the bolometer body is measured with a thermometer. The temperature difference between the non-irradiated state and the irradiated state is proportional to the power of the absorbed radiation. To relax back into its initial state the bolometer body is coupled to a thermal bath. The thermal coupling is described by the thermal coupling coefficient G. The thermal coupling coefficient governs the time in which the heat can be removed from the bolometer body.

When the absorber of the bolometer is no longer irradiated the temperature of the bolometer body T relaxes back to the temperature of the thermal bath T_b. The time dependent behavior of a bolometer that absorbs all incoming radiation can be described according to [9] using the thermal balance equation of the system:

$$P_\text{inc} = G(T - T_b) + C\frac{\mathrm{d}T}{\mathrm{d}t} \quad . \tag{2.1}$$

The speed of this temperature change or its inverse the time in which the bolometer can relaxes to its equilibrium state is defined by the ratio between the thermal coupling coefficient G and the heat capacity C of the bolometer body. This time constant can be calculated from (2.1) to

$$\tau = \frac{C}{G} \quad . \tag{2.2}$$

When the bolometer body has a large heat capacity it can store more heat than a bolometer with a small heat capacity. For a fixed thermal coupling the bolometer with the smaller heat capacity will relax much faster to the bath temperature than the bolometer with the large one. Equally for a fixed heat capacity the strength of the thermal coupling coefficient defines how fast the heat can be transferred from the

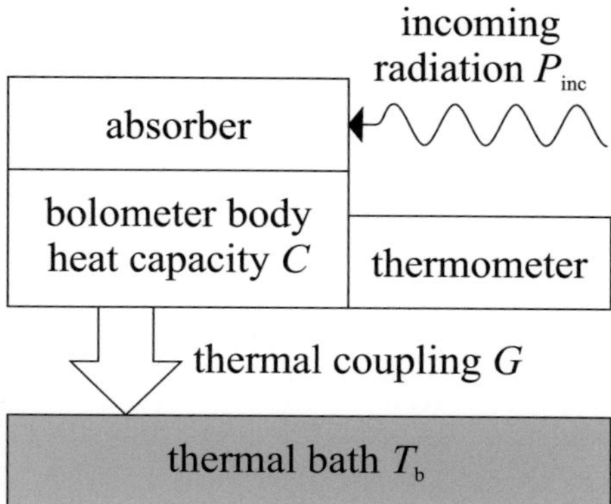

Figure 2.1: Schematic of a classical bolometer which consists of three basic parts. An absorber, the bolometer body with a heat capacity C and a thermometer. The bolometer is thermally coupled to a thermal bath with temperature T_b. The thermal coupling is described by the thermal coupling coefficient G.

bolometer body to the thermal bath. The larger the thermal coupling coefficient the faster the relaxation back into the initial state.

When a bolometer is irradiated by a radiation source with an amplitude modulated output power the temperature change of the bolometer depends on the modulation of the incoming radiation. When the power of the incoming radiation P_{Inc} consists of a portion with a constant power P_0 and a variable portion ΔP it can be written as

$$P_{Inc}(t) = P_0 + \Delta P(\omega t) \tag{2.3}$$

where ω is the modulation frequency of the output power. Using equation (2.1) and (2.3) the temperature response of the bolometer ΔT on the incoming radiation power ΔP can be calculated according to [9]

$$\frac{\Delta T}{\Delta P} = \frac{1}{G\sqrt{1+\omega^2\tau^2}} \quad , \tag{2.4}$$

where τ is taken from equation (2.2). Equation (2.4) describes the thermal sensitivity S_{th} of the bolometer to incoming radiation in [K/W]. The magnitude of the sensitivity depends on the one hand on the thermal coupling coefficient between the bolometer body and the thermal bath. On the other hand it depends through τ on the heat capacity of the bolometer body. Depending on the modulation frequency ω of the incoming radiation one of the two factors is dominant. In the limit of $(\omega\tau \gg 1)$ equation (2.4) simplifies to

$$\frac{\Delta T}{\Delta P} = \frac{1}{C\omega} \quad . \tag{2.5}$$

The bolometer response is proportional to the inverse of its heat capacity. Additionally in this operation mode the response of the bolometer becomes frequency dependent. An increase of the modulation frequency leads to a reduction in sensitivity. For high sensitivities in this mode of operation the heat capacity has to be as small as possible. In the limit of $(\omega\tau \ll 1)$ equation (2.4) simplifies to

$$\frac{\Delta T}{\Delta P} = \frac{1}{G} \quad . \tag{2.6}$$

This is the most common mode of operation for bolometers. The response increases with a reduction of the thermal coupling but the time constant of the bolometer increases. For optimal performance the factor G should be selected as small as possible. When operating the bolometer in this regime the product of the sensitivity and the speed of the bolometer is constant.

$$S_{\mathrm{th}}\tau^{-1} = const. \propto C \tag{2.7}$$

The only way to increase the temperature response of a bolometer without changing the speed is by variation of the heat capacity of the bolometer body. The heat capacity of the bolometer C can be calculated from the volumetric heat capacity c_{V} of the bolometer material and the volume of the bolometer body V to $C = c_{\mathrm{V}}V$. The volumetric heat capacity depends on the material choice for the bolometer body. The limitations for the bolometer volume are defined by the limits of the fabrication process.

The temperature response of the bolometer is directly proportional to the power of the incoming radiation. To measure this temperature change a thermometer has to be used. There are different possibilities for thermometers and depending on the type of thermometer, the response of the bolometer can be measured using different quantities. The most common form of bolometer is the resistive bolometer, where the temperature is measured via a resistor that has a well known temperature dependency $R(T)$.

When biasing a resistive bolometer with a constant current I_b the signal response is proportional to the change of incoming radiation power. It can be measured by using the voltage change due to the heating of the resistor with constant bias current

$$\Delta V = I_{\text{bias}} \Delta R \quad . \tag{2.8}$$

The electrical sensitivity S of such a resistive bolometer is defined as the change of the detector voltage ΔV over the change of radiation power ΔP. Taking equation (2.8) and (2.4) the sensitivity can be written as

$$S = \frac{\Delta V}{\Delta P} = I_{\text{bias}} \frac{\Delta R}{\Delta T} \frac{1}{G} \frac{1}{\sqrt{1 + \omega^2 \tau^2}} \quad . \tag{2.9}$$

The electrical sensitivity of the resistive bolometer has the same form as the thermal sensitivity of the bolometer and thus the same dependencies concerning C, G and τ apply. Equation (2.7) can be rewritten with the electrical sensitivity in mind to $\tau^{-1} S = const. \propto C$. This means that the sensitivity of the bolometer can still only be increased at the expense of the speed.

The introduction of a resistive thermometer gives another method to increase the sensitivity of the bolometer without decreasing its speed. The sensitivity of a resistive bolometer is directly depending on the steepness of the temperature derivative of the resistance. The temperature derivative of the resistance is a property of the material the thermometer is made of.

For the optimization of the sensitivity *and* speed of a bolometer the choice of material for the different parts of the bolometer is of great importance. The absorber should be made from a material that can absorb all of the incoming radiation. The temperature derivative of the resistance $\frac{\Delta R}{\Delta T}$ of the thermometer should be as large as possible. The heat capacity of the bolometer body should be as small as possible. And the thermal coupling of the bolometer body to the thermal bath G should be as small as possible. When taking into account all the required parameters for fast and sensitive bolometers the suitable materials are quite limited. For bolometers operated at room temperature mostly thin metal films or semiconductors are used. At room temperature the specific heat capacities of metals (e.g. platinum $c_{\text{p,pt}} = 0.13 \frac{J}{gK}$ [10]) and semiconductors (e.g. silicon $c_{\text{p,si}} = 0.75 \frac{J}{gK}$ [10]) are in the same order of magnitude so no great difference in operation should arise from this given the volume of the bolometer is similar. In

Table 2.1: *TCR* of various different materials suitable for fabrication of uncooled bolometers, fabricated with different techniques. *TCR* data taken from [11].

Technique	Material	TCR $[\mathrm{K}^{-1}]$
Sputtering	YBaCuO	2.9-3.5
CVD	$\mathrm{Si}_x\mathrm{Ge}_{1-x}$	2.4
DC sputtering + oxidation	VO_x	2.0
PLD	VO_x	2.8
Ion beam sputtering + oxidation	VO_2	2.6
RF sputtering	$\mathrm{V}_2\mathrm{O}_5/\mathrm{V}/\mathrm{V}_2\mathrm{O}_5$	2.6
RF sputtering	V-W-O	2.7-4.1
DC magnetron sputtering + annealing	VO_2	4.4
Reactive e-beam evaporation	$\mathrm{VO}_2+ \mathrm{V}_2\mathrm{O}_5$	3.2

state of the art semiconducting bolometers commonly used materials with a high temperature coefficient of the resistance ($TCR = \frac{1}{R}\frac{\Delta R}{\Delta T}$) are VaOx, SiGe or YBaCuO which range from 2 to 4%/K (see Table 2.1) or metals have a maximum TCR of 0.3%/K.

To further improve the sensitivity of a bolometer at room temperature the thermal coupling between the bolometer body and the thermal bath is the only available parameter for tuning. By increasing G higher sensitivities can be achieved. This is usually achieved by placing the bolometer on a thin membrane [12] or using a spider-web approach [13] structure. The strong thermal decoupling leads to high sensitivities in the order of $10^6\,\frac{\mathrm{V}}{\mathrm{W}}$ but also increased response times of the bolometers to several milliseconds. Materials with higher TCR at room-temperature are available but they are difficult to use as bolometers.

To reduce the temperature dependent noise contributions it is common to cool down the bolometer. Another positive side effect of the low temperature operation is the strong decrease of the heat capacity of the bolometer material with decreasing temperature. According to the Debye model [14] the phonon heat capacity of a solid material at temperatures far below the Debye temperature strongly depends on the temperature $c_\mathrm{V} \propto T^3$ where c_V is the volumetric heat capacity of the material. Therefore at low temperatures the speed of the bolometers increases significantly.

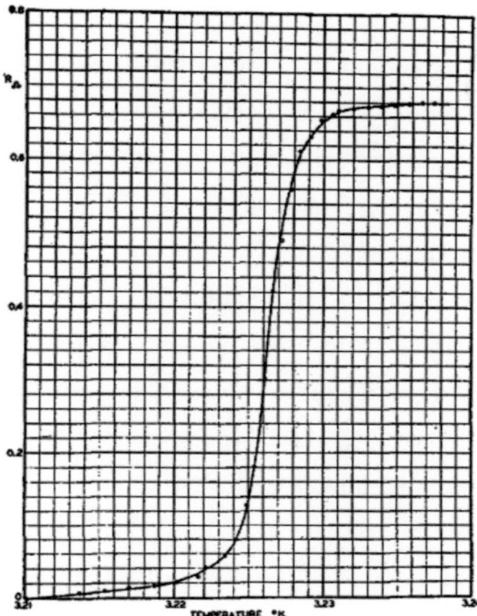

Figure 2.2: The resistance over temperature behavior of a thin superconducting tantalum wire close to its critical temperature T_C is shown. Picture taken from [15].

2.1.1 Superconducting bolometers

When operating at low temperatures superconducting materials become possible candidates as material for the bolometer body. They show a strong dependence of their resistance on the temperature at their superconducting transition temperature T_C. Such a transitions is depicted in figure 2.2. The here shown superconducting transition is from a thin tantalum wire bolometer [15]. Superconducting bolometers made from titanium have shown transition widths in the range of a few mK [16]. Due to the very small width of their superconducting transition ΔT_C the TCR of a superconductor operated on its transition can be orders of magnitude higher than for conventional semiconductor or normal metal bolometers. They are thus well suited as thermometers [17] for bolometers. A common approach to take advantage of these properties is to use composite superconducting bolometers with a non superconducting absorber and a small superconducting bolometer body as thermometer [18]. Bolometers with a superconducting thermometer have to be operated at temperatures below or close to

T_C which are in the range between a few hundred mK for aluminum or up to 80 K for YBCO. Detectors operated on the superconducting transition in direct detection mode are called transition-edge sensors (TES). This kind of detector has been proven to be very sensitive with state of the art detectors reaching a noise equivalent power of $NEP = 5 \cdot 10^{-17} \frac{\mathrm{W}}{\sqrt{\mathrm{Hz}}}$ [16] for titanium detectors at 375 mK. Though the response time of such a bolometer is as large as 13 ms.

To further increase the speed of the bolometer the heat capacity of the bolometer body has to be reduced. This can be achieved by reducing its dimensions. The heat capacity of the bolometer directly depends on its volume $C = c_V V = c_V w l d$, where w, l and d are the width, length and thickness of the bolometer element respectively. Therefore a miniaturization of the detector element is desirable.

The lateral dimensions are on the one hand limited by the technological limit of the lithography system used (which is in the range of a few 10 nm). On the other hand the superconducting properties of the thin films used for the bolometer element also depend on size and the film thickness is limited by fabrication technology for ultra-thin films to a thicknesses of a few nm. For structures that are too thin or with small lateral dimensions the superconducting transition temperature and critical current are greatly reduced (see chapter 3.1) which makes them no longer suitable for bolometer applications.

For the fabrication of fast and sensitive bolometers it can be concluded that the materials used for the thermometer/bolometer body need to have a high TCR at the operation temperature. The speed of the bolometers can be increased by deceasing the heat capacity of the bolometer. This can be achieved by reducing the volume of the bolometer body and low operation temperature. For operation at low temperatures superconducting thin film bolometers show very good performance.

2.1.2 Hot-Electron Bolometers (HEB)

The signal power of many interesting applications in the THz frequency range is very small which makes it difficult to detect. To detect such small signals there is a need for very sensitive detectors. One kind of detector that can satisfy this need is the Hot-Electron Bolometer (HEB).

When the electrons in a material can be described as thermally decoupled from the

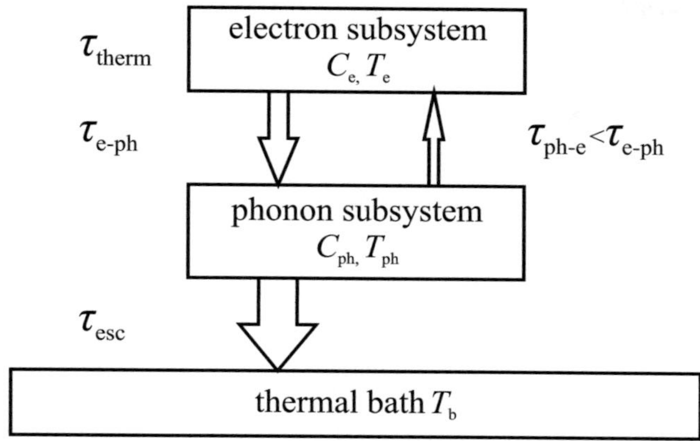

Figure 2.3: Schematic depiction of the energy transfer in a hot-electron bolometer. Taken from [22].

phonons, they can each have a different temperature. This can happen when the thermal coupling between the phonons and the electrons is weaker than the thermal coupling between the phonons and the substrate. This means that the electrons can heat up while the the phonons stay cold, thus the term hot-electron. In this situation both, the electrons and the phonons, can be in thermal equilibrium each having its own temperature where the temperature T_e of the heated electrons is higher than the phonon temperature T_{ph}. The term "hot-electron" was at first used to describe non-equilibrium electrons in semiconductors [19]. The hot-electron effect has since then been observed in metals [20] and superconducting systems [21].

In a superconducting HEB the electrons and Cooper pairs are used as the absorber and the body of the bolometer. They are also used as the thermometer. The phonon subsystem which is coupled to the thermal bath is used as the heat sink. When radiation is absorbed in such a detector the temperature increases and thus the resistance of the superconducting device changes [21]. This change of resistance is used to measure the incoming radiation.

The thermalization scheme of the electrons and phonons and their corresponding interaction times in a HEB are schematically shown in figure 2.3. $C_{e,\text{ph}}$ are the heat capacities and $T_{e,\text{ph}}$ the temperatures of the electrons and the phonons respectively. According to [22] the time for the energy transfer from the phonon to the electron sub-system $\tau_{\text{ph-e}}$ has to be much larger than the escape time τ_{esc} of phonons into the substrate for an effective decoupling of the electron subsystem from the phonon subsystem. This guarantees that the heat from the phonon subsystem is transported into the thermal bath. At the same time the electron-phonon interaction time $\tau_{\text{e-ph}}$ has to be much smaller than the phonon-electron interaction time to prevent backflow of energy from the phonon to the electron subsystem. The thermalization inside the electron and phonon subsystems is assumed to be instantaneous.

The process of detection of radiation in a HEB can be described as follows. When a photon with an energy of $\hbar\omega$ is absorbed in the HEB it breaks a Cooper pair or excites an electron and produces a highly excited electron which has almost the complete energy of the photon. The electron looses its energy by emitting high energy phonons that in turn can, with sufficient energy, break other Cooper pairs or scatter with electrons. This creates an avalanche effect of excited electrons that redistributes the energy of the absorbed photon inside the electron subsystem. Since this thermalization time τ_{therm} is much smaller than $\tau_{\text{e-ph}}$ it can be assumed that the electrons have a uniform temperature T_{e}. This thermal energy can then, on a timescale of $\tau_{\text{e-ph}}$, be transferred to the phonon subsystem which is at a temperature of T_{ph}. The stored heat is completely removed out of the system after a time τ_{esc} when it is transferred to the thermal bath with a temperature T_{b}. The time constant corresponding for the energy backflow from the phonons to the electrons is depending on the ratio of the respective heat capacities and can be taken from equilibrium conditions ($\tau_{\text{ph-e}} = \tau_{\text{e-ph}} \frac{C_{\text{p}}}{C_{\text{e}}}$). The heat capacities in turn have a strong dependence on the detector temperature. The heat capacity of the electrons is inversely proportional to the temperature $c_{\text{e}} \propto T^{-1}$ while the phonon heat capacity below the Debye temperature follows a power of three law $c_{\text{ph}} \propto T^{-3}$ [14]. At too low temperatures the electronic heat capacity can become larger than the phonon heat capacity and the hot-electron effect seizes to exist.

The time constants for the different interactions between phonons, electrons and the substrate are material and temperature dependent. To achieve low electron-electron interaction times, which are in part responsible for the thermalization inside the electron subsystem, it is preferred to have a disordered material and temperatures which are as high as possible for the material to encourage scattering processes. The electron-electron interaction time can be written according to [23] as $\tau_{\text{e-e}} \propto \frac{1}{TR_{\square}} ln^{-1}(R_{\square})$. So,

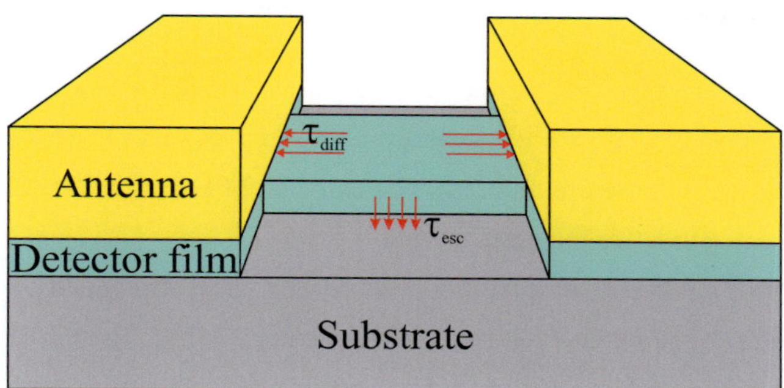

Figure 2.4: Sketch of the detector area of a typical Hot-Electron Bolometer. It consists of a thin detector film with thickness d and an antenna structure. The detector element with length l and width W is defined between the edges of the antenna structure. Heat generated in the detector element can escape by diffusion (τ_{diff}) or by thermal escape into the substrate (τ_{esc}) as indicated by the red arrows.

high operation temperatures and a large square resistance $R_\square$ of the film are required for a fast thermalization.

The electron-phonon interaction time is depending on the strength of the interaction between phonons and electrons. It has been shown in [24] that the temperature dependence of this interaction follows $\tau_{\mathrm{e\text{-}ph}} \propto T^{-m}$ where m is a experimentally obtained parameter which varies between 1 and 3 for different materials. The intrinsic electron-phonon interaction time in niobium nitride [23] has been found to have a temperature dependency of $\tau_{\mathrm{e\text{-}ph}} \propto T^{-1.6}$. This temperature dependency implies that HEBs exhibit shorter relaxation times at higher temperatures. Since an operation at higher temperatures is improving the speed of the HEB, materials with high T_C are better suited for use in HEB mixers.

Design of a Hot-Electron Bolometer

To remove the heat from the phonon sub-system before it can flow back to the electron sub-system the phonon escape time has to be small. For a given material it greatly depends on the geometry of the fabricated bolometer. Figure 2.4 shows the layout of the detection area of a typical Hot-Electron Bolometer. The HEB consists of a thin superconducting detector film with a gold layer for an antenna structure on top. The detector element is defined between the large metallic contacts used for an antenna structure for radiation coupling and a RF readout line.

The dimensions of the detector element are chosen so that its DC resistance is close to that of the antenna structure with an impedance of 74 Ω [25] and the impedance of the coplanar waveguide for the readout which is 50 Ω.

The geometry of the detector is then defined by the sheet resistance of the thin film times the number of squares N required for the detector bridge to reach 50 Ω ($NR_\square \approx$ 50Ω). For a sheet resistance between 100 and 500 Ω, which is typical for niobium nitride, N is in a range of $N = 0.1 \cdots 0.3$. The width W of the detector bridge is chosen in a way that the smallest dimension (usually the length l) is in the range of a few hundred nanometers. Depending on the length of the device the heat has two different paths on which it can leave the system indicated by the red arrows in figure 2.4.

Hot electrons can diffuse out of the bolometer into the large cold metallic contacts when they are close enough. This happens on the time scale of the diffusion time τ_{diff}. The electrons can also scatter with the lattice of the bolometer and cool down by emission of phonons into the substrate which happens on the scale of the escape time τ_{esc}. Both of these cooling mechanisms are present in every bolometer but depending on the size and the material one of the two is dominant. Detectors can thus be separated into two categories, diffusion-cooled [26] and phonon-cooled bolometers [27].

For the diffusion to be the dominant process the length of the detector must be shorter than the thermal diffusion length [26]

$$L_{\mathrm{diff}} = \sqrt{D\tau_{\text{e-ph}}} \tag{2.10}$$

where D is the diffusion constant for the electrons in the material used as detector. This cooling mechanism is mostly relevant in superconducting thin-film hot-electron bolometers fabricated from clean and well ordered material like niobium with diffusion coefficients of $D \geq 1\,\mathrm{cm/s^2}$ [28]. Due to their large diffusion constant the critical length of the detector is in the range of a few hundred nanometers [29] which is easy to fabricate with state of the art lithography and deposition systems. The relaxation time $\tau_{\mathrm{diff}} \approx (l/4)^2/D$ in these systems is depending on the length l of the detector bridge and independent of the temperature [30]. The drawback is that due to the high diffusion coefficients the influence of the contacts on the detector element increases.

This leads to a strong influence of the normal metal contacts on the superconducting thin film of the detector. This so called proximity effect (see chapter 3.3.1) may lead to an unpredictable behavior of the bolometer. When fabricating bolometers from disordered materials like niobium nitride one has to account for their low diffusion coefficient. With a diffusion coefficient of $D = 0.5 \text{ cm}^2/\text{s}$ [31] and a short electron-phonon interaction time in the order of ten picoseconds [32] the required bolometer length for diffusion cooling is in the range of 10 nanometers. This size is not suitable for fabrication. In this case the limiting time constant for the removal of the heat from the bolometer body is the thermal escape time from the thin film into the substrate (see equation (2.14)). When the escape time becomes too large the heat gets trapped inside the phonon subsystem. It was found by Kaplan that the probability of phonons escaping from a thin film is depending on several factors [33]. For thin films and if specular and diffuse scattering of the phonons at the film substrate interface is assumed the escape time can be approximated to

$$\tau_{\text{esc}} = \frac{4d}{\alpha u} \tag{2.11}$$

where d is the thickness of the superconducting thin film, α the transmission coefficient for phonons from the film to the substrate and u the speed of sound in the thin film. The transmission coefficient depends on the difference of the speed of sound in the thin film and the substrate material and their respective densities. By choosing a substrate with a perfect matching of the phonon spectra the escape time can be greatly reduced.

Since the escape time limits the speed of our detectors for very fast detectors the films have to be made as thin as possible and the phonon transmission coefficient between the thin film and the substrate should be as high as possible.

2.1.3 Theoretical model of the HEB

To describe the response of a hot-electron bolometer to incoming radiation theoretically the two-temperature model [21] was developed. It describes the time evolution of the temperature of the bolometer by introducing the time constants mentioned earlier in this chapter. It is assumed that the thermalization of the electrons and the phonons in their respective subsystems happens instantaneous over the whole detector and a uniform temperature distribution can be assumed. The system can be described by the coupled heat balance equations of the respective sub-systems. These equations are normally non linear but become linear in the case of a small signal excitation and close to T_C. Using these assumptions the thermal balance equation for the electron sub-system can be written according to [22] as

$$C_e \frac{dT_e}{dt} = P_{inc} - \frac{C_e}{\tau_{e\text{-}ph}}(T_e - T_{ph}) \tag{2.12}$$

and for the phonon sub-system as

$$C_{ph} \frac{dT_{ph}}{dt} = \frac{C_{ph}}{\tau_{ph\text{-}e}}(T_e - T_{ph}) - \frac{C_{ph}}{\tau_{esc}}(T_{ph} - T_b) \tag{2.13}$$

where $C_{e,ph}$ are the heat capacities and $T_{e,ph}$ the temperatures of the electrons and the phonons respectively. The phonon-electron scattering time constant $\tau_{ph\text{-}e} = \frac{C_{ph}}{C_e}\tau_{e\text{-}ph}$ was taken from equilibrium conditions. The energy of an electron is completely removed from the system when it has gone through the phonon subsystem and to the substrate. The effective energy relaxation time of the electrons can be described according to [22] as

$$\tau_{e,av} = \tau_{e\text{-}ph} + \left(1 + \frac{C_e}{C_{ph}}\right)\tau_{esc} \tag{2.14}$$

From equation (2.14) the dependence of the relaxation time of the HEB on the different parameters can be derived. For a $\frac{C_e}{C_{ph}} \ll 1$ the response is dominated by the larger of the two time constants τ_{esc} and $\tau_{e\text{-}ph}$. For an increasing ratio the escape time becomes the dominant term and is mainly responsible for the thermal decay. This model can be used to simulate the response of the detector to, for example, pulsed radiation and explains the pulse shape and the decay times seen in such experiments (see chapter 5). This model does not take into account the dissipated power from the DC bias current. Also the assumption of a uniform temperature distribution over the complete device is not sufficient to explain the operation of a HEB in mixing mode. Therefore the hot-spot model was developed [34]. This model describes the bolometer by a one dimensional thermal balance equation where the temperature of the electrons varies along the bridge. It is assumed that the HEB is pumped with a local oscillator and that the absorbed power of the local oscillator is uniformly absorbed over the whole bridge. The second assumption is that the complete DC power is dissipated only in a small hot-spot in the center of the detector element [35]. In the hot-spot the electron temperature is above the critical temperature T_C and the superconductivity is suppressed. The heat-balance equations for the HEB inside of the hot-spot can be written according to [34] as

$$-K\frac{\mathrm{d}^2 T}{\mathrm{d}x^2} + \frac{C_e}{\tau_{\text{e-ph}}}(T_e - T_b) = j^2\rho + P_{\text{LO}} \qquad (2.15)$$

and outside of the hot-spot as

$$-K\frac{\mathrm{d}^2 T}{\mathrm{d}x^2} + \frac{C_e}{\tau_{\text{e-ph}}}(T_e - T_b) = P_{\text{LO}} \qquad (2.16)$$

where K is the thermal conductivity in the bolometer thin film, j is the current density of the bias current flowing through the bridge, ρ is the resistivity of the bolometer thin film in the normal state and P_{LO} is the absorbed local oscillator power. When the incoming radiation power changes, the size of the hot-spot increases and decreases accordingly [36]. This change of the hot-spot size results in a resistance change over the HEB and can be measured in form of a voltage/current change.

To predict the behavior of the HEB to a change in incoming radiation the model assumes a uniform absorption of the local oscillator radiation over the whole device. In real devices this assumption is not correct. Disturbances of the superconducting order parameter at the edges of the bridge due to the metallic contacts in the finished detector can lead to formation of different hot-spots at the edges of the bolometer. If the quality of these contacts is not well defined this may lead to an irreproducibility of detector characteristics between different detectors and to additional noise and RF losses in the bolometer due to uncontrolled interface resistances in these areas. Therefore the quality of the contacts and the interfaces has to be controlled for optimal performance of the detector.

The hot-spot model can be used to describe the behavior of a HEB mixer qualitatively very well for radiation frequencies in the THz region. When the radiation frequency of the local oscillator is only several hundreds of GHz it has been observed that the predictions made by the hot-spot model do not fit the experimental results. It has thus been proposed that the radiation incoming on the bolometer directly interacts with vortex/anti-vortex pairs inside the detector element [37]. This results in a increased intermediate frequency bandwidth of the HEB when operated at lower frequencies. This has to be taken into account when characterizing mixers at low local oscillator frequencies.

2.2 Direct detection of radiation

When used for direct detection the HEB is operated as a classical bolometric detector and reacts directly to the incoming radiation. The detector does not differentiate between different frequencies and polarizations. It acts as a power detector and only integrates over the incoming radiation power. Thus the detector is not sensitive to the phase of the incoming radiation.

To limit the spectrum of the detected radiation passive filters are used to filter out the desired spectrum. Frequency dependent information and information depending on the polarization of the radiation has to be gathered from the radiation by application of external polarizers and spectrometers often in the form of tunable interferometers.

2.3 Heterodyne detection of high frequency radiation

To obtain the spectral information of the measured radiation without the need for complex filters and interferometers detectors can be operated in the heterodyne detection mode. In heterodyne detection the incoming radiation signal is mixed down with a local oscillator signal in a mixer element (here the HEB) to frequencies easily accessible by standard readout electronics. The accessible spectral ranges are depending on the frequencies of the local oscillator and the intermediate frequency bandwidth of the mixer as explained later in this section. The advantage of fast superconducting HEB mixers is that they can have an intermediate frequency bandwidth of several GHz in the THz frequency range [26] which allows efficient measurements of large spectral ranges. When a HEB is operated as a heterodyne detector it is kept at a temperature below the critical temperature of the detector element. The detector is biased with a bias current I_b and pumped to a resistive state with a local oscillator with a power of P_{LO}. This leads to the formation of a hot-spot inside the bolometer as described by equation (2.15) and (2.16). When radiation of a frequency close to the LO frequency is coupled into the detector the length of the hot-spot changes with the frequency difference between LO and signal frequency.

One of the main advantages of HEB mixers is that they require only few nW of pumping power to operate which is perfect for the THz region where strong and compact

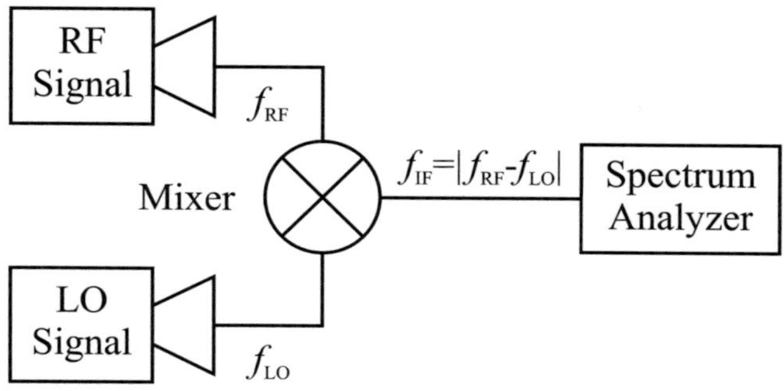

Figure 2.5: Schematic of a heterodyne detection setup. The incoming radiation (f_{RF}) is mixed with the radiation of a local oscillator (f_{LO}). The output signal of the mixer has a frequency of $f_{IF} = |f_{RF} - f_{LO}|$

sources are not readily available. Also the noise of superconducting HEB detectors is very low (see e.g. [38]).

The underlying principle of the heterodyne detection is schematically shown in Figure 2.5. The detection setup uses a source signal with a frequency f_{RF} that has to be analyzed, and a local oscillator signal at a frequency f_{LO} close to the signal frequency. By mixing the source signal with the local oscillator signal the mixer product of the two frequencies lies in an intermediate frequency (IF) range accessible by modern electronics $f_{IF} = |f_{RF} - f_{LO}|$. The intermediate frequency bandwidth of the heterodyne detection is limited by the relaxation time of the detector element. The signal output from the detector can then be amplified and measured. For the characterization either a spectrum analyzer or a dedicated bandpass filter with a power meter attached to it can be used.

To describe the response of the detector to incoming radiation we first assume that the incoming radiation and local oscillator radiation are of sinusoidal form. This radiation is absorbed in the bolometer and then induces an oscillating voltage with equivalent frequency. The induced voltage in the detector can be written as

$$V(t) = V_{RF}cos(\omega_{RF}t) + V_{LO}cos(\omega_{LO}t) \tag{2.17}$$

where V_{RF} is the voltage induced by the radiation signal and V_{LO} is the voltage induced by the local oscillator. ω_{RF} and ω_{LO} are the corresponding frequencies of the radiation.

The power dissipated in the bolometer can then be calculated to

$$P(t) = \frac{V^2}{R} = (V_{RF}cos(\omega_{RF}t) + V_{LO}cos(\omega_{LO}t))^2$$

$$= \frac{1}{R}\left(\frac{V_{RF}^2}{2}(1+cos(2\omega_{RF}t)) + \frac{V_{LO}^2}{2}(1+cos(2\omega_{LO}t))\right) \tag{2.18}$$

$$+ V_{RF}V_{LO}(cos((\omega_{RF}+\omega_{LO})t) + cos((\omega_{RF}-\omega_{LO})t)))$$

The time constants of the HEB detector are in the order of a few picoseconds. Thus the detector can not follow high frequency changes with several hundred GHz up to several THz used for the LO and the RF signal, $\omega_{RF,LO}$. Therefore the first three terms in equation (2.18) give a constant contribution to the dissipated power. The remaining time dependent part is the intermediate frequency contribution from the $(\omega_{RF}-\omega_{LO})$ term. Substituting $P_{RF/LO} = \frac{V_{RF/LO}^2}{R}$ we can calculate the time dependent part of (2.18) to

$$P(t) \propto \sqrt{P_{RF}P_{LO}}cos((\omega_{IF})t) \quad . \tag{2.19}$$

with $\omega_{IF} = \omega_{RF} - \omega_{LO}$ the intermediate frequency. Since the cosine is symmetric around zero the same output signal can be created by two different input signals. The absolute value of $|\omega_{RF} - \omega_{LO}|$ gives us the output frequency of the mixer. This leads to an operation mode called double sideband (DSB) mode. When one of the sidebands is suppressed through the use of filters or similar devices the operation is called single sideband (SSB) mode.

2.4 Antenna-coupled superconducting bolometers

To couple the radiation from the free space to the detector element there are several possibilities. The most common method to transfer a free-space wave into a current is an antenna. For the purpose of high integration directly on chip a planar antenna element is preferred.

Depending on the requirements of the experiment different kinds of planar antenna structures can be used to couple the radiation. There were two kinds of different antennas used in this work desinged by our group (for more in-depth information see [39]). The double-slot antenna and the log-spiral antenna. The double-slot antenna (see figure 2.6 on the left side) is a resonant antenna with a linear polarization and a

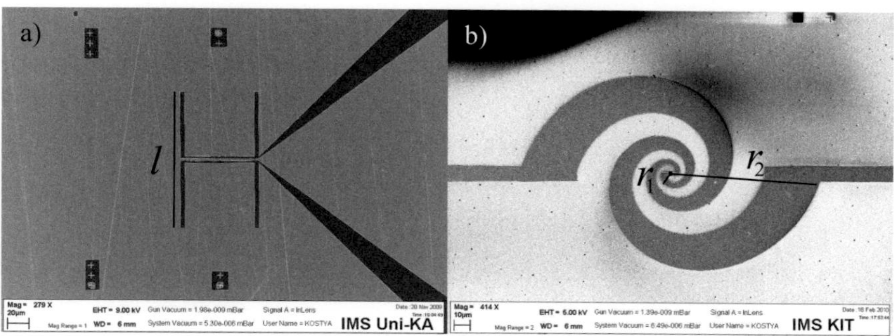

Figure 2.6: SEM images of different antenna types. In (a) a double-slot antenna is depicted. The center frequency of this antenna is defined by the length of the slots of the Antenna. The image (b) shows a log spiral antenna where the minimum and maximum frequencies are defined by r_1 and r_2 respectively [40].

small bandwidth. The center frequency of the double slot antenna is defined by the length of the arms which correspond to a quarter wavelength. This kind of antenna has a bandwidth of only up to 30 % and the to be detected radiation frequencies have to be in this narrow bandwidth of the antenna.

The log-spiral antenna (Figure 2.6 on the right side) is a traveling wave antenna and has a very weak polarization dependency. The minimum frequency of the antenna is defined by the inner radius of the spiral structure and the maximum frequency is designed by the maximum radius of the antenna element.

For all antenna designs the detection element is placed inside a gap in the center of the antenna element and has typical dimensions of $W \times L$ of 1 μm$\times$0.3 μm.

The material from which the antennas are fabricated has to have a very low resistance to reduce losses in the antenna structure. It should also not directly absorb the radiation power and convert it to thermal energy. The material of choice for the antenna structure is a gold layer with several hundred nanometers of thickness which has a low resistivity and almost no direct absorption of radiation happens up to several THz.

2.5 Noise in Bolometers

Hot-electron bolometers are used for measurement of very weak energy signals. The ultimate limiting factor to signal measurement is the noise generated by the receiver system and the background noise of the incoming radiation. When the noise signal is larger than the signal of the incoming radiation no measurements are possible. Thus for the measurement of small signals the reduction of the noise is of great importance.

2.5.1 Noise equivalent power

When operated in direct detection mode the noise equivalent power (NEP) is often used as a figure of merit to characterize the noise performance of detector systems. In the case of the HEB detector one can differentiate between several different contributions to the NEP. Using the NEP it is possible to add and compare the different contributions of noise to the total. The NEP is defined as the radiation power P_{inc} incident on the detector which produces an output signal of the magnitude of the noise P_{noise} of the detector in a bandwidth of one Hz. The total NEP in the detector can be calculated by adding the squares of the different NEP which arise from uncorrelated noise sources [41].

The noise in the detector has various sources [9] and the main contributors for the noise in an ideal bolometer can be identified as the photon noise from the background radiation, the Johnson noise and the phonon noise.

The total NEP and the Johnson noise of the bolometer can according to [9] be written as

$$(NEP)^2 = (NEP)^2_{\mathrm{Johnson}} + (NEP)^2_{\mathrm{phonon}} + (NEP)^2_{\mathrm{photon}} \quad . \qquad (2.20)$$

The Johnson noise is given due to random fluctuation of the electrons inside the bolometer and has a value of

$$(NEP)^2_{\mathrm{Johnson}} = (4kTR)/|S|^2 \qquad (2.21)$$

with k the Boltzmann constant, T the detector temperature, R the resistance of the bolometer and S the sensitivity of the bolometer. The noise contribution from the

Johnson noise can be reduced by increasing the sensitivity of the detector, by reducing the resistance or the operation temperature. In the case of a niobium nitride hot-electron bolometer the resistance is fixed due to the need of matching to the antenna structure and the operation temperature is fixed near T_C. The only possible way to reduce the Johnson noise contribution in this case is to increase the sensitivity of the detection element which can be achieved by methods mentioned earlier in this chapter e.g. reduction of the bolometer volume or use of high TCR materials. The phonon noise term

$$(NEP)^2_{\text{phonon}} = 4kT^2G/\eta \tag{2.22}$$

with G the coupling between detector thin film and substrate and η the absorptivity of the bolometer, stems from the thermal fluctuations inside the bolometer. It is caused by the phonons passing from the bolometer over the thermal conductance G to the substrate and is calculated from the temperature fluctuations inside the system. It can be suppressed by reducing the thermal coupling of the bolometer to the thermal bath which is also desired for an increased sensitivity.

The NEP contribution from the photon noise has its source in the background radiation incoming on the detector. Usually bolometric systems use filters with a transmittance $\tau(v)$ to block the incoming radiation power in the unwanted frequency ranges. A portion of the background radiation is still incoming to the detector and the noise of the background radiation adds to the total noise of the system. This noise contribution is not dependent on the bolometer fabrication but is a characteristic of the incoming radiation source and its temperature. Thus detectors only limited by this noise contribution are called background limited detectors.

The above mentioned quantities for the NEP are all calculated for a system which is in a thermal equilibrium. It has been shown [42] that by calculating the noise from non-equilibrium theories these quantities can be further reduced. The main contributor of noise in real detectors is the phonon noise. To reduce this quantity the thermal coupling to the film and the quality of the interfaces through which this thermal coupling happens has to be improved.

Since it is difficult to experimentally access the different parts of equation (2.20) directly an often taken figure of merit is the noise temperature of the detection element. This value corresponds to the NEP in a way that $T_N = NEP/k\sqrt{B}$ where B is the measurement bandwidth of the system. It is most commonly used when comparing different superconducting mixers.

2.5.2 Noise temperature

To compare the performance of different mixer devices in the THz range the noise temperature T_{noise} is the most common used figure of merit. To directly measure the noise temperature of a mixer setup the detector is pumped into its operation point by a local oscillator. Then the mixer is exposed to the thermal radiation coming from a hot load (usually a black body radiator at room temperature) and a cold load (usually a black body radiator cooled down to 77 K using liquid nitrogen). The signal response of the detector to this hot an cold load is measured. Using the Y-factor method the noise temperature of the system can be directly calculated.

To calculate the Y-factor of a detector it is irradiated with two radiation sources with a well known temperature and radiation power like the hot/cold load described above. Using the following formalism according to [43] the noise temperature can be calculated. The output power of the detector will then be

$$P_{\text{out,hot/cold}} = G(P_{\text{in,hot/cold}} + P_{\text{noise}}) \tag{2.23}$$

where G is the gain of the detector, $P_{\text{in,hot/cold}}$ is the radiation power of the hot/cold load and P_{noise} the noise of the detector in the bandwidth B of the system. The Y-factor is the ratio between the output power of the detector under hot and cold load $Y = \frac{P_{\text{hot}}}{P_{\text{cold}}}$. Combining this formula with equation (2.23) the Y-factor can be calculated to

$$Y = \frac{P_{\text{in,hot}} + P_{\text{noise}}}{P_{\text{in,cold}} + P_{\text{noise}}} \quad . \tag{2.24}$$

Solving this equation for P_{noise} the noise power can then in turn be calculated to

$$P_{\text{noise}} = \frac{P_{\text{in,hot}} - Y P_{\text{in,cold}}}{Y - 1} \tag{2.25}$$

For radiation detectors the hot and cold load used to irradiate the detector is a black body radiator at liquid nitrogen temperatures $T_{\text{cold}} = 77$ K and one at room temperature $T_{\text{cold}} \approx 300$ K. The noise temperature can then be easily calculated to

$$T_{\text{noise}} = \frac{T_{\text{hot}} - Y T_{\text{cold}}}{Y - 1} \tag{2.26}$$

using the Rayleigh-Jeans law $P = kT$ where k is the Boltzmann constant. This is valid for devices characterized at very high temperatures where the Rayleigh-Jeans law is still valid. Kerr suggested [43] that at operation temperatures of the devices close to absolute zero at high radiation frequencies and with devices close to the quantum noise limit the physical temperatures in formula (2.26) have to be corrected, to account for this discrepancy. The physical temperatures $T_{\text{hot,cold}}$ have to be replaced by the Callen and Welton definition [44]

$$T_{\text{C\&W}} = T \left(\frac{\frac{hf}{kT}}{exp\left[\frac{hf}{KT}\right] - 1} \right) + \frac{hf}{2k} \tag{2.27}$$

in order to derive the noise temperature which accounts for zero fluctuation noise $\frac{hf}{2k}$.

2.6 Bolometers for the THz frequency range

As has been discussed in the previous sections the choice of material has a very large influence on the properties of the HEB. There are only a few materials which can be used to fabricate fast and sensitive THz detectors for the THz range [45]. Niobium nitride has been used widely for bolometric detectors in the THz range over the last decades. It is one of few materials that enables fabrication of phonon cooled HEB mixers with large intermediate frequency bandwidth of several GHz. Due to the low diffusion coefficient of the niobium nitride in the order of $D = 0.5cm^2/s$ [31] and short intrinsic electron-phonon interaction time the dominant cooling process is phonon cooling. With the ability to fabricate high quality thin films with a thickness of a few nanometers while still retaining high T_C a low escape time, and thus a high intermediate frequency bandwidth, can be achieved. Because of this we use on niobium nitride as superconducting thin film for the detectors in this work.

2.7 Requirements and challenges for HEB detectors for the THz spectrum

To fabricate fast and low noise hot-electron bolometers for the THz frequency range the design of the hot-electron bolometer devices has to be improved and optimized. Several parameters of the bolometer design are already fixed due to technological and experimental requirements. The operation temperature of the hot-electron bolometer is set above liquid helium temperatures. Additionally application requires that the speed of the detector or respectively the intermediate frequency bandwidth when operated as mixer should be at least a few GHz. The desired frequency range dictates that an antenna structure has to be used to couple the radiation into the detector element.

The design for hot-electron bolometers consist of a multilayer structure. In figure 2.4 the different layers, namely the superconducting thin film layer and the antenna layer on top, can be seen. For a better adhesion of the antenna layer on the detector film an additional layer has to be interposed between the two. This is done *in-situ* during the deposition of the antenna. The length and width of the detector element have to be chosen in a way that the resistance of the detector element provides good matching resistance to the antenna impedance.

The combination of these requirement already limits the choice of the superconducting material for the detector thin film. The choice of the superconducting material also determines the minimum film thickness which can be used for the detector layer which in turn defines the relaxation time of the bolometer according to equation (2.11). As described in Chapter 2.6 niobium nitride is the most suitable material for a phonon-cooled hot-electron bolometer and is used for the detectors fabricated in this work. Optimization of the superconducting niobium nitride thin film for the detector layer (see chapter 3.1) is thus the first step in developing optimized hot-electron bolometers. To reduce the noise introduced to the bolometer due to the antenna contacts (see chapter 2.1.3) the superconducting and normal conducting properties of each of the bolometer layers have to be analyzed separately (see chapter 3.2). Additionally the influence of multiple layers on each other have to be examined and taken into account. Lastly, and most importantly the interfaces between the different layers have to be examined and taken into account when fabricating multilayer structures for hot-electron bolometer detectors.

Another aspect of the bolometer optimization is the thermal coupling of the superconducting thin film to the substrate. As discussed in chapter 2.1 this is a key parameter for the speed of the detector element. Depending on the requirements of the experiment it can be desirable to reduce the thermal coupling between the detector element and the thermal bath to increase the sensitivity of the device according to equation

(2.9). To reduce the thermal coupling of the bolometer to the substrate several different approaches have been used like spiderweb or membrane structures. In this thesis a novel approach is presented in Chapter 4 where the thermal coupling is reduced by placing the superconducting thin film on a elevated mesa structure.

3 Interfaces and proximity effect in multi-layer structures of THz HEB devices

To improve the performance of hot-electron bolometers, one has to acquire a deep understanding of the different components that a HEB consists of and how their properties influence each other. In figure 3.1 the schematic layout of a typical bolometer is depicted. It consists of four layers that are: the substrate and the detector film which is fabricated from ultra-thin ($\approx$ 5 nm) niobium nitride. On top of the detector layer a buffer layer is deposited on the ultra-thin niobium nitride film. This layer provides an electrical contact and improves the adhesion of the final layer of gold.

The interfaces between these four layers have a strong influence on the properties of the neighboring layers and thus on the operation of the bolometer. The interface between the detector film and the substrate determines the escape of phonons from the niobium nitride film to the substrate which is the key parameter for the speed of a direct detector (see equation (2.2)) and determines the intermediate frequency bandwidth in case of a HEB mixer [46][47]. The quality of the surface of the substrate and its lattice matching to the niobium nitride also determines the crystalline and superconducting properties of the niobium nitride thin films [48].

Fabrication of the antenna requires that the gold layer on top of the niobium nitride is deposited *ex-situ*. This leads to an exposure of the niobium nitride film to the surrounding atmosphere and thus a contamination of the thin film surface which would introduce a high contact resistance between the gold and the niobium nitride. By depositing an *in-situ* normal metal buffer layer the contact resistance is greatly reduced. Yet the usage of a normal metal buffer layer results in a partial suppression of superconductivity in the region of the detection element due to the direct contact of a normal conductor and a superconductor. In turn this leads to a non-homogeneous Hot-Spot formation in the detection element, which was shown in [49], and leads to a strong degradation of device performance. It has been proposed [50] that precleaning of the

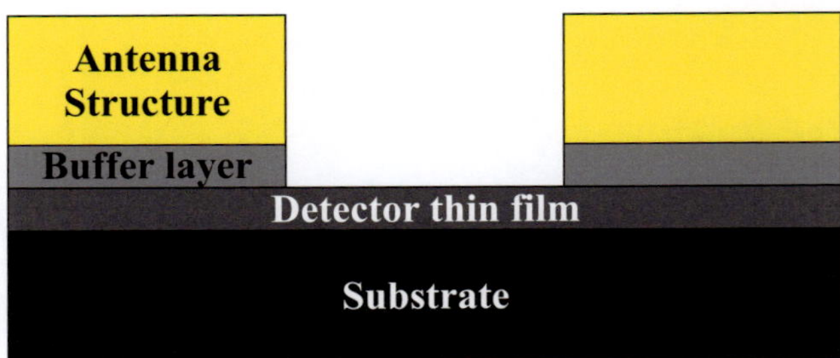

Figure 3.1: The detector consist of three different layers which are deposited on the substrate. The first layer is the active detector film layer, the second layer is a buffer layer which serves as a electrical and mechanical stabilization layer between the Antenna structure and the detector film.

interface and application of a superconducting buffer layer reduce these effects by increasing the phonon transparency between the buffer layer and the detector film.

The goal of the optimization of the interfaces between the different layers of the HEB structure is, on the one hand, an increase of the intermediate frequency bandwidth. This can be achieved through thinner films and a good matching between the substrate and the thin film. On the other hand, a strong reduction in noise temperature due to a more uniform formation of the hot-spot can be achieved by the application of superconducting buffer layers and reduction of the interface resistance between antenna and detector.

In this chapter we will first present results on the development of ultra-thin film deposition technology for the niobium nitride detector films and the influence of the substrate film interface on the superconducting properties (see chapter 3.1). The critical temperature in respect to the film thickness will be analyzed in the light of the proximity effect model. The dependence of the superconducting and normal conducting properties of two different materials (niobium and niobium nitride) on the film thickness and different fabrication conditions will be analyzed and their fitness for the buffer layer will be discussed in chapter 3.2.

The respective bi-layers niobium/gold and niobium nitride/gold, the interface between the gold and the buffer layer and the proximity effect between these films will be discussed in chapter 3.3.

In chapter 3.4 the influence of the interfaces in the multilayer will be examined and the performance of a HEB fabricated with an optimized multilayer structure will be presented.

3.1 Ultra-thin superconducting NbN films for the detector layer

The key element in a superconducting hot-electron bolometer detector is the superconducting thin film used for the detector element (see figure 3.1). In the detectors fabricated in this work the material of choice for the ultra-thin films is niobium nitride due to its general suitability for detectors in the THz frequency range (see chapter 2.7) and widely accepted material for HEB [51][52]. For optimal performance of the HEB the fabrication process has to be optimized for small film thicknesses while still retaining high critical temperatures.

3.1.1 Influence of the substrate and deposition temperature

The superconducting properties of niobium nitride vary greatly with deposition conditions and substrate properties. Therefore an analysis of the superconducting properties of ultra-thin niobium nitride films deposited on different substrates and under different conditions was conducted. The thin films were either deposited on two-side polished single-crystalline (100) oriented silicon substrates which were fabricated using the floating-zone method or on single side polished R-cut sapphire substrate.

The base pressure inside the deposition chamber was in the range of 1.5×10^{-7}mbar. The base pressure was created by a turbo-molecular pump and the chamber was baked out overnight to remove adsorbed water in the chamber walls. Before deposition, the niobium target was cleaned by sputtering off its surface in a pure argon atmosphere at a pressure $P_{Ar} = 3 \times 10^{-3}$mbar for about 10 min. Then nitrogen gas was added to the atmosphere with a partial pressure of $P_{N2} = 2 \times 10^{-4}$mbar. After stabilization of the discharge voltage the deposition was made. The thickness d of the superconducting thin films was calculated from the deposition time. The deposition rate was evaluated from the results of a high resolution transmission electron microscopy [48] study of similar niobium nitride films deposited onto Si substrate and was found to be 10 nm/min.

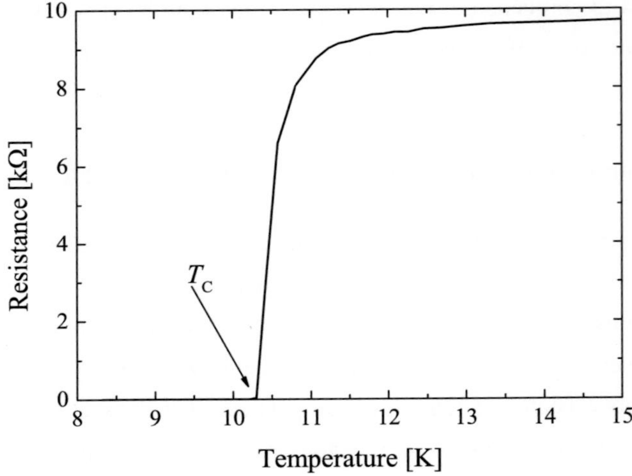

Figure 3.2: Superconducting transition of a 13 nm thick niobium nitride film deposited on Silicon substrate. The critical temperature T_C is indicated by an arrow at the point where the resistance first drops to zero.

Niobium nitride was deposited with reactive DC-magnetron sputtering at substrate temperatures of 700° C for silicon substrate and 750° C for sapphire substrate. Silicon and sapphire have been chosen as substrate materials due to their high transparency in the THz frequency range and their good matching to the lattice of the niobium nitride. R-cut sapphire has a lattice constant of $a_\mathrm{Al2O3} = 4.7\text{Å}$ and silicon has a lattice constant of $a_\mathrm{Si} = 5.4\text{Å}$ while the niobium nitride lattice constant is $a_\mathrm{NbN} = 4.4\text{Å}$. Right after deposition the temperature dependence of the film resistance was measured in the temperature range from 4.2 K to room temperature by four-probe technique and the zero critical temperature T_C was measured. Figure 3.2 shows an example curve for the measuring of the critical temperature. The resistance over temperature for a 13 nm thick niobium nitride film close to its superconducting transition is shown. We define the critical temperature as the point where the resistance first drops to zero as indicated by the arrow.   The thickness dependence of the zero resistance critical temperature is shown in figure 3.3 for three different deposition conditions. The black squares correspond to niobium nitride thin films deposited on a sapphire substrate heated to 750° C during deposition, the red circles correspond to niobium nitride deposited on

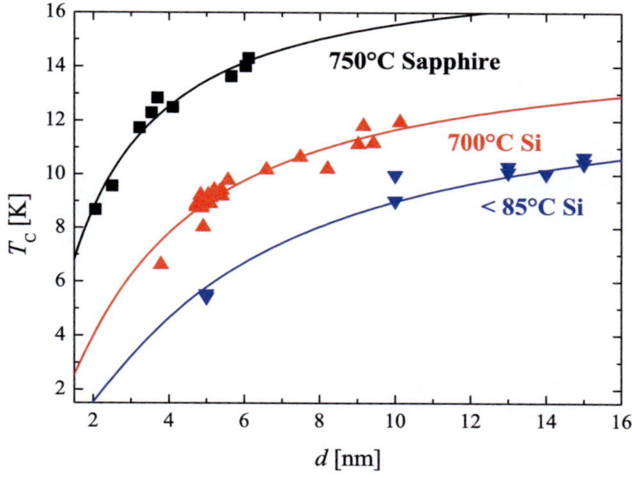

Figure 3.3: Superconducting transition temperature T_C of niobium nitride thin-films over thickness for reactively sputtered films. From top to bottom T_C on heated sapphire (Al_2O_3) ($T_{dep} = 750°$ C, black ■), heated silicon ($T_{dep} = 700°$ C, red ▲) and silicon at room ($T_{dep} \leq 85°$ C, blue ▼) is depicted.

silicon heated to 700° C during deposition and the blue triangles correspond to niobium nitride deposited on silicon at ambient temperature. A general feature that all three curves in figure 3.3 have in common is a strong decrease of their critical temperature for thinner films. This is a well known behavior for thin superconducting films and will be explained in detail in chapter 3.1.2. Figure 3.3 also shows a strong dependence of T_C on the substrate properties. Since the R-cut sapphire lattice has a better lattice matching to the lattice of niobium nitride than the silicon we expect T_C to be higher on the sapphire substrate which is indeed shown for the films deposited on heated sapphire substrate. They show a critical temperature which is more than 2 K higher than that of the films deposited on heated silicon substrate. The strong difference between the hot deposited and cold deposited niobium nitride on silicon can be explained by a much higher mobility of the atoms at elevated temperatures which in turn leads to an improved crystalline growth while the films grown at ambient temperature are amorphous. This leads to a difference in the critical temperature between the hot deposited and the cold deposited niobium nitride of about 2 K.

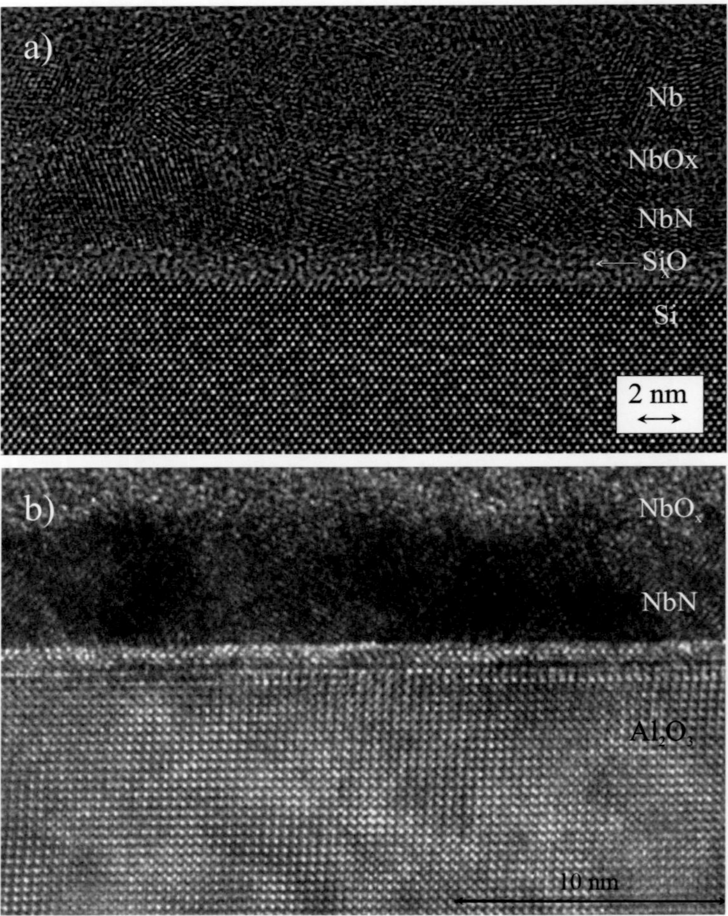

Figure 3.4: Transmission electron microscopy images of niobium nitride deposited by reactive magnetron sputtering on heated silicon (a) and heated sapphire (b). The niobium nitride on silicon shows a polycrystalline growth with grain sizes of several nanometers. The niobium nitride on sapphire shows a quasi epitaxial growth with grain sizes in the range of the film thickness.[53]

To analyze the influence of the substrate properties on the growth of the niobium nitride thin films, transmission electron microscopy analysis of niobium nitride films deposited on heated substrates were made. Figure 3.4 shows transmission electron microscope (TEM) images taken on niobium nitride films deposited on heated silicon (a) side and those deposited on heated sapphire (b). The niobium nitride deposited on heated silicon substrate was covered with an *ex-situ* deposited layer of niobium. The niobium nitride shows a polycrystalline growth with grain sizes in the range of a few nanometers in this sample. The grains are tilted relative to each other and show no preferred orientation. On the top and the bottom of the niobium nitride layer clearly amorphous regions are visible where the substrate is covered with native oxide or the deposited niobium nitride layer was exposed to the atmosphere respectively. We have analyzed the elemental distribution in these regions using electron-energy loss spectroscopy (EELS) [53]. It was found that there are large amounts of oxygen in the interface regions. In between the oxide layers the growth of the niobium nitride is stoichiometric with no oxygen included. The niobium films on top of the niobium nitride film also shows a polycrystalline growth but in the interface region there seems to be an amorphous layer. In these regions oxides of niobium are formed which are not superconducting. This reduces the effective thickness of the niobium nitride film reduces the critical temperature T_C. Another factor influencing the critical temperature is the amorphous layer of silicon oxides that hamper the epitaxial growth of the niobium nitride.

The niobium nitride grown on heated sapphire substrate is shown in figure 3.4 b). We can see that there is no amorphous layer and thus no formation of niobium oxides on the surface of the substrate. This leads to an epitaxial growth of nano-crystallites which are limited in height by the film thickness but show a preferred orientation of the grown crystalline structure. On top of the niobium nitride film again a amorphous layer of niobium oxides can be seen which reduces the effective superconductor thickness. In conclusion, the lattice mismatch of niobium nitride with a lattice constant of $a_{NbN} = 4.4\mathring{A}$ on Si substrate with $a_{Si} = 5.4\mathring{A}$ is larger than that on R-plane sapphire substrate with a lattice constant of $a_{Al2O3} = 4.7\mathring{A}$. The niobium nitride grows on the silicon in a polycrystalline shape with small crystallites while epitaxial growth can be observed on the sapphire. The critical temperature of the niobium nitride films deposited on silicon substrate is suppressed due to surface effects at the grain boundaries inside the film. In niobium nitride on sapphire these boundaries are almost non-existent due to the epitaxial growth. This leads to a large difference in T_C of about 2.5 K for 5 nm thin films deposited on heated substrates.

3.1.2 Proximity effect in thin superconducting films

As discussed in chapter 2.1.2 the superconducting thin films for the detector element should be as thin as possible to reduce the heat capacity of the detector and increase its speed (see equation (2.2)). As it turns out a reduction of the film thickness leads to a decrease of the critical temperature of the film. This effect is attributed to the proximity effect first proposed by Cooper [54]. All films in Figure 3.3 show a strong dependence of their critical temperature on the film thickness. This effect has already been observed in a lot of different superconducting materials [55]. Cooper proposed that the interaction constant of a superconductor $[N(0)V]$ is greatly reduced when it comes into contact with a normal metal. For the most simple case, where the electrons in the superconducting and the normal conducting film have the same effective mass, the same Fermi velocities and a perfect interface between the metal and the superconductor the new interaction constant of the S+N bi-layer is according to [54]

$$[N(0)V]_{\text{S+N}} = \frac{d_{\text{S}}}{d_{\text{N}} + d_{\text{S}}} [N(0)V]_{\text{S}} \quad , \tag{3.1}$$

where d_{S} and d_{N} are the thicknesses of the superconductor and the normal metal respectively. Due to the exponential dependence between the superconducting transition temperature and the interaction constant [54] a very thin normal conducting film is sufficient to influence the superconducting properties:

$$T_C \propto e^{\frac{1}{N(0)V}} \tag{3.2}$$

Combining equation (3.1) and equation (3.2) the superconducting transition temperature of a superconductor/metal film can be easily calculated

$$T_C(d_{\text{S}}) = T_C(\infty)e^{-\frac{d_{\text{N}}}{d_{\text{S}}N(0)V}} \tag{3.3}$$

where $T_C(\infty)$ is the superconducting transition temperature of the bulk material.

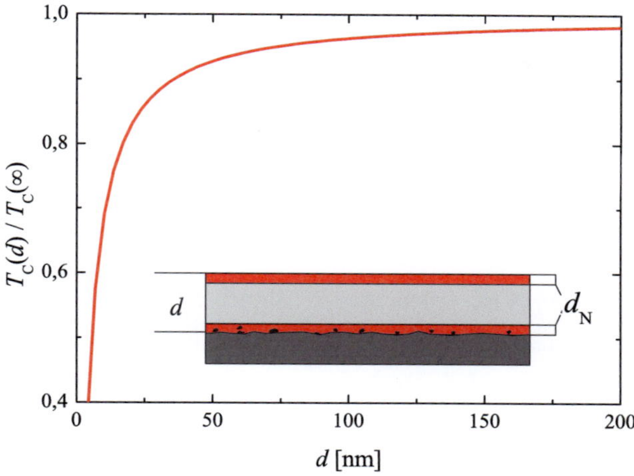

Figure 3.5: Calculation of the normalized critical temperature over thickness for a superconducting thin film in contact with a layer with destroyed superconductivity of thickness d_N. The contact between the superconducting film and the normal conducting film leads to a suppression of the superconducting transition temperature proposed by [54].

Even though this model assumes a very special case of two almost identical materials it can be used to describe the behavior of thin superconducting films from one material [55]. Due to oxidation at the surface the superconductivity in a thin layer with a combined thickness of d_N at the surfaces of the film gets destroyed and we get normal conducting layers. We assume this layer of destroyed superconductivity to be constant in thickness since the formation of the surface layer should be independent of the superconducting film thickness. The influence of this surface layer grows with a decrease of the film thickness. The resulting sandwich of a superconducting film between layers of suppressed superconductivity can be described by a N/S structure and applying Coopers theory (see equation (3.3)) we can see a clear decrease in the transition temperature. The behavior of the transition temperature of a superconducting film with layers of suppressed superconductivity with a combined thickness d_N is calculated in figure 3.5.

When applying this theory to the measurement data of the niobium nitride films we can calculate the value of d_N and $T_\mathrm{C}(\infty)$ for the different substrate materials and deposition

Table 3.1: d_N and $T_C(\infty)$ values for the three different niobium nitride deposition conditions

	$T_C(\infty)$ [K]	d_N [nm]
Hot deposited NbN on sapphire	17.9	0.5
Hot deposited NbN on silicon	15.2	0.8
Cold deposited NbN on silicon	13.9	1.4

conditions. The results of this calculation are shown in table 3.1. The small values for d_N mean that only a few unit cells thick normal conducting films are responsible for the decrease in superconductivity which fits to our assumption of a thin surface layer that is responsible for this effect.

The values in table 3.1 show that the layer of destroyed superconductivity is the smallest for the hot deposited niobium nitride films on sapphire substrate which is of about half the thickness as for the hot deposited silicon. This seems reasonable since the TEM images in figure 3.4 show a amorphous niobium oxide layer on the top and the bottom of the niobium nitride film deposited on silicon. On sapphire only on the top of the niobium nitride film an amorphous layer is present. The thicknesses of the niobium oxide layers in the TEM is in the range of a few nanometers which is substantially larger than the values calculated from equation (3.3). This is probably due to the very strong simplifications of the used model. Nevertheless, it enables us to predict the transition temperature of our thin films very well. It also reinforces the idea that the removal of the surface oxide of the silicon is crucial to achieve higher T_C values for these films. To further increase the superconducting properties of the thin films one has to be aware of these facts. It has to be taken care that the interface between the substrate and the deposited films is free of oxides and that the substrate has a good lattice matching to the niobium nitride.

3.2 Superconducting thin films as buffer layer

As discussed in chapter 2.1.2 and shown in figure 3.1 the detector element is embedded between large metal contacts for an antenna structure. For the detectors fabricated in this work the chosen antenna types were planar antennas. The planar antennas might differ in their form depending on the experimental requirements (see chapter 2.4) but always consist of a thick gold layer which is deposited on top of the detector film (see figure 3.1). Since the adhesion of gold *ex-situ* deposited on niobium nitride is very poor, a buffer layer is introduced between the antenna and the detector film.

The fabrication process of the detector requires that the antenna and the buffer layer are patterned using lift-off technology. Thus the niobium and gold layer have to be deposited at ambient temperature to prevent overheating of the used resist. The most common buffer layer materials used are chromium ($T_C = 3$K) and titanium ($T_C = 0.4$K) [50]. They are normal metals at the operation temperature of the HEB devices ($T = 4.2$K). This results in a suppression of the superconductivity in the detector film due to the proximity effect. To prevent such an influence superconducting materials like niobium or niobium nitride can be used as materials for the buffer layer.

The detector element has a critical temperature of about 10 K. To reduce the influence of the proximity effect on the critical temperature of the detector element the buffer layer should have a critical temperature which is similar or even higher than that of the detector element. At the same time the thickness of the buffer layer should be as small as possible to reduce the resistive losses in this part of the detector. Therefore the superconducting properties of niobium and niobium nitride thin films are measured and the suitability of these materials as buffer layers is evaluated.

3.2.1 Room temperature deposition of superconducting thin films

The niobium or niobium nitride films were deposited on one side polished low resistive silicon wafers. These wafers were cut prior to deposition to pieces 10 by 10 mm in size. Structures were created via lift-off technique and the patterning of the resist was done by photo lithography prior to deposition. The films were then DC magnetron sputtered at ambient temperature. The sputtering chamber is equipped with a 3 cm diameter Ar ion-beam gun, which is used for pre-cleaning of the substrate, a 2 inch magnetron with a niobium target for sputtering of niobium and reactive sputtering of niobium nitride films. The chamber is also equipped with a 2 inch magnetron with a gold target for the *in-situ* deposition of Au directly on top of the superconducting thin film without the use of a normal metal buffer layer. The base pressure of the main chamber is in the range of 10^{-7} mbar, which is reached by pumping with a cryopump.

Samples were sputtered on a rotating sample holder to ensure homogeneity of the thickness of the thin films. The niobium thin films were deposited with a deposition rate of 0.26 nm/s at an argon pressure of $5 \cdot 10^{-3}$ mbar. For the niobium nitride films a mixture of argon and nitrogen was used. The argon pressure was kept at $5 \cdot 10^{-3}$ mbar and a nitrogen partial pressure of $2.4 \cdot 10^{-3}$ mbar was used. This results in a deposition rate of 0.2 nm/s.

The results from chapter 3.1 show that there is an amorphous native oxide on the silicon substrate. This layer is influencing the growth of the thin films and thus their critical temperature and normal conducting properties. To analyze the influence of the native oxide layer on the substrate surfece on the critical temperature the substrate was pre-cleaned for some of the deposited films. For the *in-situ* pre-cleaning prior to the deposition of the superconductor an Ar ion-beam-gun with an energy of 100 eV was used. This gives an etching rate of the silicon of less than 0.3 nm/min. Cleaning of the substrates was done for two minutes. After deposition the superconducting and normal state properties of the samples were characterized with a four probe measurement setup. The critical temperature in these measurements was defined as the point where the resistance of the films dropped below 0.1% of the normal state resistance R_N of the structures for the first time.

3.2.2 Superconducting and normal conducting properties of niobium and niobium nitride thin films

The dependence of the critical temperature on thickness for niobium films deposited on substrates that were prepared with and without pre-cleaning is shown in figure 3.6 (a). A strong decrease of the superconducting transition temperature for film thicknesses of less than 50 nm can be seen for both kind of samples as expected according to the proximity effect theory in chapter 3.1.2. The samples deposited with pre-cleaning show an improved transition temperature which is about 0.3 K higher than for the uncleaned samples. The lines in the graph are a best fit according to equation (3.3). Using this equation a value of $T_\mathrm{C}(\infty) = 9.5 K$ was calculated. Figure 3.6 (b) shows the residual resistivity ratio ($RRR = R_{300}/R_{10}$) of the respective niobium films. The residual resistivity ratio is a rough indicator for the crystalline quality and the purity of the film. It is also reduced in thinner films due to enhanced scattering effects at the

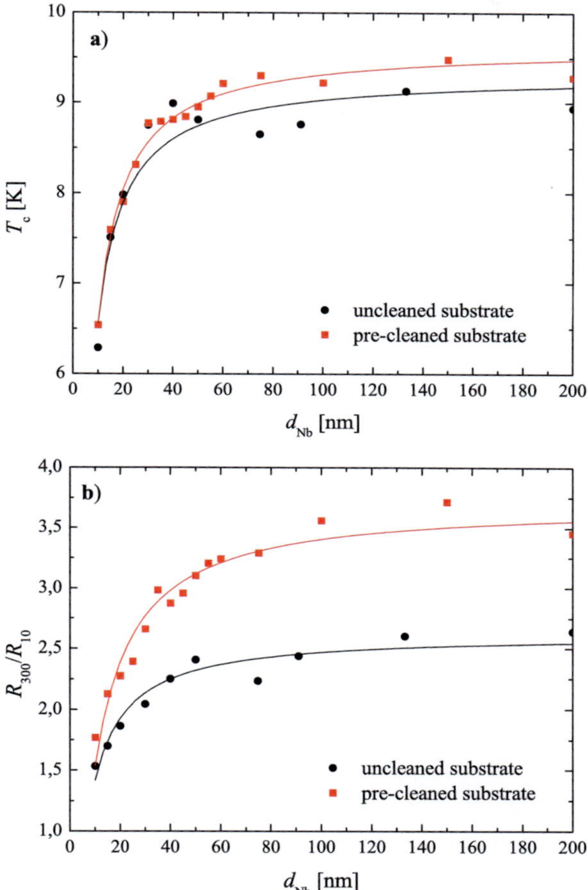

Figure 3.6: Superconducting transition temperature over film thickness of niobium films sputtered on room temperature silicon substrate (a) and the corresponding residual resistivity ratio of the thin-films (b). The black line corresponds to a deposition made on unmodified silicon substrate. The red line corresponds to samples where the silicon substrate was *in-situ* pre-cleaned by means of Ar-ion milling before deposition. The lines in graph (a) are best fits according to equation (3.3). The lines in graph (b) are to guide the eye.

Table 3.2: Calculated superconducting transition temperature and thickness of the layer with destroyed superconductivity according to equation (3.3).

	$T_C(\infty)$ [K]	d_N [nm]
Nb	9.6	1.2
NbN	13.2	1.2

film surface. This behavior was predicted for thin metallic films [56] where the mean free path in the bulk material is in the same range as the film thickness. The strong difference between the *RRR* for uncleaned films (*RRR* ≈ 2.5) and pre-cleaned films (*RRR* ≈ 3.5) indicates that removing the amorphous oxide layer from the substrate surface leads to formation of a more homogeneous niobium film.

The $T_C(d)$ dependence of the niobium nitride layer deposited on substrates prepared with or without pre-cleaning can be seen in figure 3.7 (a). The same reduction of the critical temperature for decreasing film thickness as for the niobium thin films can be seen in the experimental data. As before this can be attributed to the proximity effect. A strong decrease of the critical temperature can again be seen for thin films below 25 nm and the $T_C(\infty)$ was calculated to $T_C(\infty) = 13.1\,\text{K}$.

The dependence of RRR for the niobium nitride films can be seen in figure 3.7 (b). The influence of the pre-cleaning is not as pronounced as in the case of niobium. Niobium nitride is a chemically stable compound material and is not as sensitive to the availability of oxygen on the substrate surface as the niobium films. The *RRR* values of the niobium nitride films are less then unity for the whole thickness range. Having a *RRR* of below unity for all thicknesses implies that the resistivity is governed by scattering at grain boundaries and impurities inside the film. The pre-cleaning has almost no bearing on the RRR value.

In conclusion, niobium and niobium nitride are both suitable candidates for a buffer layer of a hot-electron bolometer. The superconducting transition temperature of both materials is above 8 K for films with a thickness larger than 10 nm for niobium nitride and about 25 nm for niobium respectively (see figure 3.8). Using equation (3.3) values for $T_C(\infty)$ and d_N were extracted from the measured critical temperature data for both films and are presented in table 3.2. The values for d_N are equal for the two materials. Both materials have good adhesion to the surface after deposition and strong getter

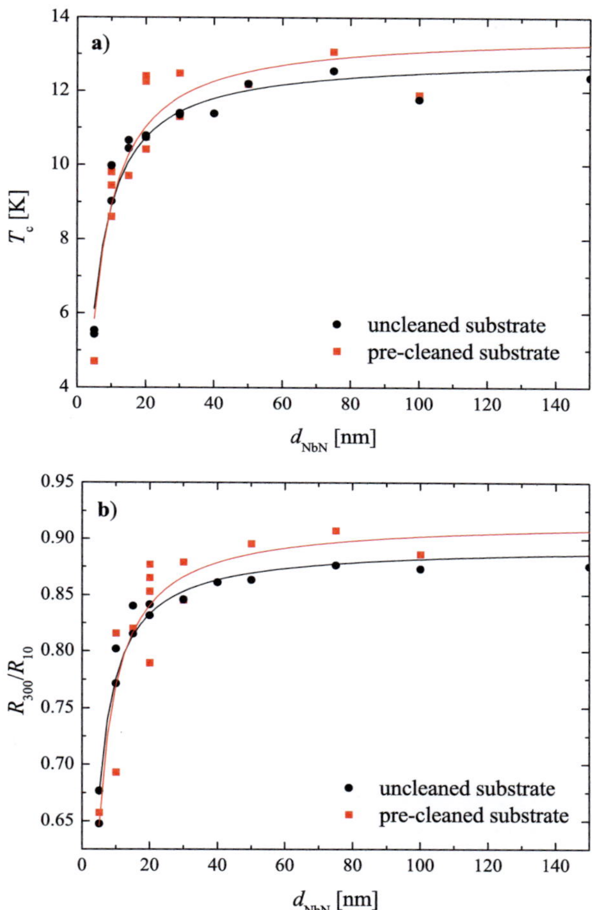

Figure 3.7: Superconducting transition temperature over film thickness of niobium nitride films sputtered on room temperature silicon substrate (a) and the corresponding residual resistivity ratio of the thin-films (b). The black line corresponds to a deposition made on unmodified silicon substrate. The red line corresponds to samples where the silicon substrate was *in-situ* pre-cleaned by means of Ar-ion milling before deposition. The lines in graph (a) are best fits according to equation (3.3). The lines in graph (b) are to guide the eye.

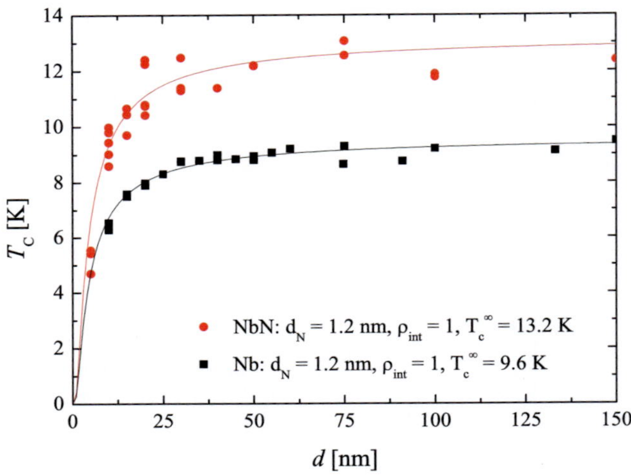

Figure 3.8: Comparison of T_C of Niobium and Niobium Nitride films depending on thickness. niobium nitride has over the whole thickness range a much higher T_C than the niobium film. Since we want the buffer layer in our detector to be as thin as possible but still achieve a T_C of ever 9 K niobium nitride is identified as more suitable buffer layer material. The lines are best fits according to Equation (3.3).

material qualities. This enables both films to be used as replacement for the normally used normal metal titanium or chromium buffer materials. Additionally the high critical temperatures is a promising indicator that it will be possible to create bi-layer structures of niobium or niobium nitride and gold which will have a critical temperature above the critical temperature of the detector film. Niobium nitride with its larger critical temperature of is the more promising candidate of these both. The next section will take a closer look on the superconducting properties of these bi-layers.

3.3 Superconducting-normal metal bi-layers for antennas

As described in the previous section the HEB detector has to be embedded into an antenna structure to be sensitive to THz radiation. The material of choice for the antenna structure is gold (see chapter 2.4). In the standard bolometer designs an additional titanium or chromium layer is added between the gold and the niobium nitride thin film to increase the adhesion. Since the antenna structure consists of normal metal it will influence the superconducting properties of the detector layer by means of the proximity effect when deposited on top of the detector film [57]. It was shown that the noise temperature and the IF bandwidth of a HEB can be greatly improved when this interface is taken into account [58]. To reduce the influence of the proximity effect between the antenna and the detector element a superconducting buffer layer (see chapter 3.2) is used.

To analyze the behavior of such superconductor/normal metal bi-layers, films fabricated from niobium and niobium nitride as discussed in section 3.2 were analyzed with a gold layer deposited on top. To fabricate these films the niobium or niobium nitride films were prepared as described in the previous chapter. After deposition of the superconducting film an *in-situ* gold layer was deposited immediately on top. The *in-situ* deposition is necessary to prevent contamination of the interface between the two films to keep it as clean as possible. By adding a normal metal layer with a thickness d_N on top of the superconducting thin film the superconductivity in these structures gets substantially suppressed. In figure 3.9 the superconducting transition of an uncovered niobium nitride film with a thickness of $d_{NbN} = 20$ nm and an uncovered niobium film with the same thickness are depicted. Also identical films covered with a gold layer with a thickness of $d_{Au} = 280$ nm are presented. The resistance of these films is normalized to their value just above the transition for better visibility of the transition regions. The superconducting transition of the covered film has been reduced by 2.4 K for niobium nitride thin films and by 1.8 K for niobium films. The reduction of the critical temperature in these bi-layers is due to the proximity effect.

To be able to characterize the interface between the gold film and the superconducting thin film the transition temperature of the superconductor with varying thicknesses of gold on top have been analyzed. Therefore samples with a fixed thickness of the superconducting film were deposited and an *in-situ* gold layer with a thickness between 10 and 200 nm was deposited on top. Afterwards the resistance over temperature was measured and the critical temperature was evaluated. Figure 3.10 shows the critical temperature of the bi-layer over the thickness of the gold d_N that was deposited on top. Niobium films with a thickness of 20 nm show a very strong dependence of their critical temperature on the thickness of the gold layer. It drops by $\approx 30\%$ for gold

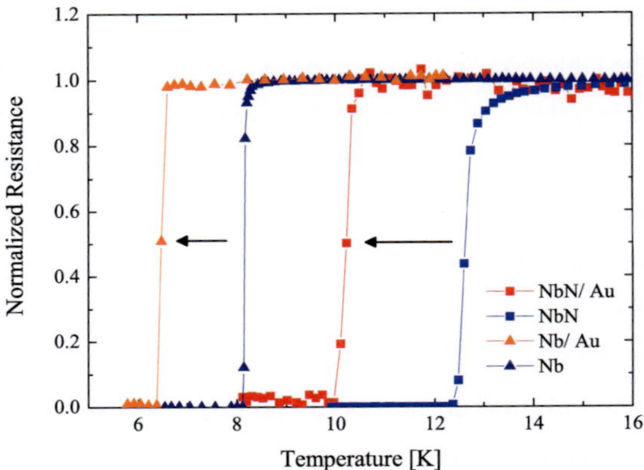

Figure 3.9: Superconducting transition of a single niobium nitride (blue), niobium (dark blue) (20 nm) layer and a niobium nitride/gold (red), niobium/gold (orange) (20 nm/280 nm) bi-layer normalized to their resistance at 10 K. The proximity effect reduces the transition temperature by 2.4 K for the niobium nitride/gold bi-layer and by 1.8 K for the niobium/gold bi-layer.

layers with a thickness of over 100 nm to a value of $T_C = 5.4K$. Films with a thickness of 33 nm on the other hand show only a very weak dependence on the thickness of the gold layer. The T_C of these films drops only $\approx 5\%$ to a value of $T_C = 8K$. This result was expected since the relative thickness of the two layers is relevant for the suppression of the superconductivity. For the niobium nitride/gold bi-layer structures a strong influence of the thickness of the gold layer on the underlying niobium nitride layer with a thickness of 18 nm can also be observed (see figure 3.11). By increasing the gold layer thicknesses to over 100 nm a strong decrease of the critical temperature by 30% down to $T_C = 8.4K$ can be observed. By further increasing of the thickness of the niobium nitride film up to 20 nm the superconducting transition temperature of a bi-layer with a 280 nm thick gold layer on top can be increased up to 10 K (see figure 3.9). This exceeds the transition temperature for ultra-thin hot-deposited niobium nitride films for the detector element. This makes this bi-layer a perfect candidate for the usage as an antenna layer.

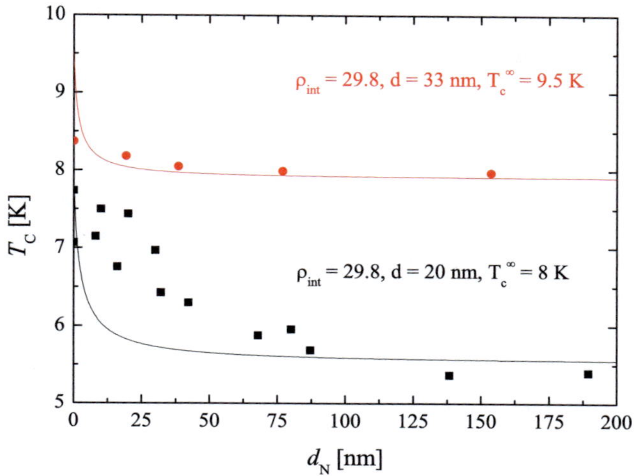

Figure 3.10: Dependence of the critical temperature T_C on thickness of the gold layer d_N for two different film thicknesses of niobium. The red dots correspond to the transition temperature of a 33 nm thick niobium film and the black squares to a 20 nm thick film respectively. The solid lines are calculated using equation (3.4).

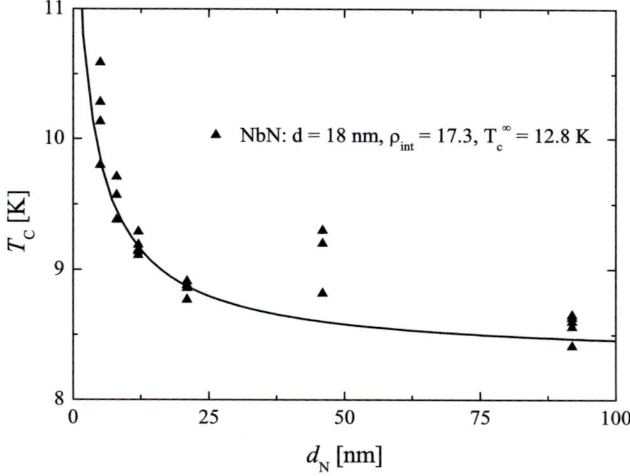

Figure 3.11: Transition temperature T_C over the thickness of the top gold layer for a niobium nitride thin film with a thickness of 18 nm. The black curve is a fit according to (3.4). The calculated interface resistance amounts to $\rho_{int} = 17.3$.

3.3.1 Proximity effect in superconductor-normal metal bi-layers with non-perfect interface

The behavior of the critical temperature of thin superconducting films can be described by the proximity effect discussed in chapter 3.1.2 and the superconducting transition temperature of S/N bi-layers can be calculated with equation (3.3). Since this model assumes a perfect interface between the superconducting film and the normal conducting film and similar Fermi velocities the theory is limited to the description of simple systems. Fominov and Feigel'man [59] took into account the different Fermi velocities of the thin films and also introduced a interface resistance to account for imperfections in the interface between the films. This led to the following formula:

$$
\ln\frac{T_C(\infty)}{T_C} = \frac{\tau_N}{\tau_S + \tau_N}\left[\Psi\left(\frac{1}{2} + \frac{\tau_S + \tau_N}{2\pi T_C \tau_S \tau_N}\right) - \Psi\left(\frac{1}{2}\right) \right.
$$
$$
\left. -\ln\sqrt{1 + \left(\frac{\tau_S + \tau_N}{\tau_S \tau_N \omega_D}\right)^2} \right] \quad , \tag{3.4}
$$

where $T_C(\infty)$ is the transition temperature of a film with infinite thickness, ω_D is the Debye-frequency, Ψ is the di-gamma function, and τ_N and τ_S are defined by

$$
\tau_N = 2\pi\frac{v_N d_N}{v_S^2}\rho_{int}, \quad \tau_S = 2\pi\frac{d_S}{v_S}\rho_{int} \quad . \tag{3.5}
$$

The values $v_{N,S}$ and $d_{N,S}$ are the Fermi velocities and the thicknesses of the superconducting and normal metal layers, respectively. The dimensionless parameter ρ_{int} represents the magnitude of the resistance at the interface R_{int} per conduction channel and is given in units of the quantum resistance R_q.

$$
\rho_{int} = R_{int}\frac{2N_{ch}}{R_q} \tag{3.6}
$$

The interface resistance depends on the number of conduction channels N_{ch} available which is not an easily accessible experimental parameter.

The strength with which the superconductivity is suppressed depends on the thickness of the normal metal layer and the interface resistance ρ_{int}. The theory developed in [59] is used to describe this effect. In figure 3.12 the influence of different values of the interface resistance have been plotted. A clear dependency on the quality of the

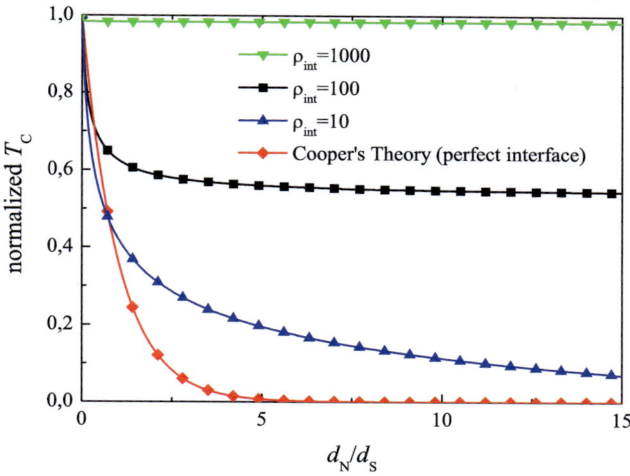

Figure 3.12: Superconducting transition temperature of a bi-layer normalized to the T_C of a single film over $\frac{d_N}{d_S}$ where d_S is the thickness of the superconducting layer and d_N the thickness of the normal metal layer. Curves were calculated with (3.4) for different values of the interface resistance ρ_{int}.

interface is visible. For a non transparent interface with a very high interface resistance there is no suppression of the superconducting transition temperature. When the transparency of the interface increases and with that the interface resistance decreases a decrease of the superconducting transition temperature for large $\frac{d_N}{d_S}$ is visible. The superconductivity in a thin superconducting film can be almost completely suppressed when the interface between the two films is perfectly transparent ($\rho_{int} = 1$) and the thickness ratio between the normal conducting and the superconducting film $\frac{d_N}{d_S}$ is very large.

Applying the above theory to the experimental data we showed earlier in this chapter we obtain the solid lines in the figures 3.10 and 3.11 for niobium and niobium nitride respectively. From these lines we can extract the values of the interface resistance of the two different setups. As can be seen when looking at the critical temperature in figure 3.10 the experimental data and the theoretical calculations do not fit very well for small film thicknesses. The suppression at low gold film thicknesses is not as large as expected from the theory. A possible explanation is a increased contact resistance

ρ_{int} and a reduced influence of the proximity effect. The origin of this increase of the contact resistance might be attributed to formation of oxide layers on top of the niobium layer inside the vacuum chamber. To estimate the interface resistance between the superconductor and the normal metal thin film the critical temperatures of bi-layers with a thickness of $d_{Au} > 75$ nm were used for fitting. The so-calculated interface resistance ρ_{int} for the niobium films amounts to $\rho_{int,Nb} = 29.8$.

For the niobium nitride films we have a very good agreement between the theory calculation and the experimental data as can be seen in figure 3.11. The interface resistance between the gold and the niobium nitride can be calculated to $\rho_{int,NbN} = 17.3$. This is only about 60% of the interface resistance between the niobium and the gold layer.

The strong difference indicates that the interface between niobium nitride and gold must be more transparent than between niobium and gold. This might be explained by the higher reactivity of niobium compared to niobium nitride. The higher reactivity might already lead to a contamination of the surface of the niobium films inside the deposition chamber due to residual gases. Even though the chamber has a very low base pressure of less than $p_{base} \approx 1 \cdot 10^{-7}$ mbar and very pure argon is used as process gas some residual oxygen might leak into the chamber during processing and contaminate the surface.

In conclusion, the examined films show very promising results in regard to application in THz detectors. We have measured a relatively low resistivity of $\rho_{Nb} \approx 1 \cdot 10^{-7} \Omega m$ for niobium thin films. But the critical temperature of the Nb/Au bi-layer structure needs a superconducting buffer layer with a thickness of at least 30 nm to have comparable T_C values as the detector element.

For niobium nitride thin films we measured a resistivity of $\rho_{NbN} \approx 4 \cdot 10^{-6} \Omega m$ which is more than an order of magnitude larger than the resistivity of the niobium thin films. Due to their innate higher critical temperature buffer layers from niobium nitride can be made as thin as 15 to 20 nm. This still results in a niobium nitride/gold bi-layer which has a critical temperature which is higher than that of the detector element. The interface resistance ρ_{int} of the bi-layers is small for both materials (see table 3.3). Compared to the values in [59] which are in the range of $\rho_{int} \geq 80$ both interfaces seem much better. In the light of the unclear difference between the theoretical and

Table 3.3: Calculated superconducting properties and interface resistance (see equation (3.4)) of niobium(niobium nitride)/gold bi-layers on Si substrate.

	$T_C(0)$ [K]	ρ_{int}
$Nb(20\ nm)$	8	29.8
$Nb(33\ nm)$	9.5	29.8
$NbN(18\ nm)$	12.5	17.3

experimental results for the niobium layer and the much higher ρ_{int} niobium seems to be not the best choice as buffer layer material. The niobium nitride/gold films show only about 60% of the interface resistance of the niobium/gold films which may be attributed to the reduced reactivity of the niobium nitride layer to residual gases in the deposition chamber during deposition. This and the conclusions from chapter 3.3 mark niobium nitride as the best possible buffer layer material for superconducting HEBs.

3.4 Detector fabrication with superconducting buffer layer

Applying the results of this chapter to the detector fabrication a detector with an optimized antenna structure was fabricated and characterized [60]. The detector layer was fabricated with the process described in chapter 3.1 on a heated silicon substrate (see figure 3.1). The deposited niobium nitride film had a thickness of 6 nm and a critical temperature of 9.5 K after patterning.

Afterwards the detector was patterned using e-beam lithography and the antenna structure was deposited on the detector film according to the process developed in chapter 3.3. The antenna bi-layer consisted of a 20 nm thick niobium nitride buffer layer with a 280 nm thick *in-situ* gold layer on top. To improve the superconducting properties of the antenna bi-layer and the detector film a pre-cleaning by means of Ar ion milling was conducted before the deposition.

3.4.1 Superconducting transition of an optimized detector element

The resistance-over-temperature curve for the fabricated detector can be seen in figure 3.13 indicated by the black dots. The critical temperature of the structured buffer

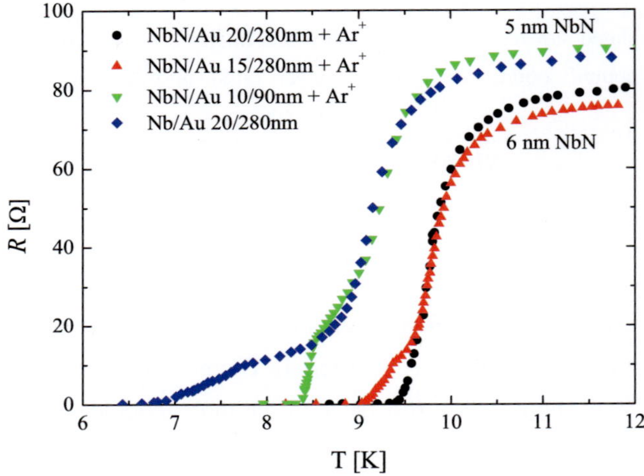

Figure 3.13: Superconducting transition of HEB detectors fabricated with different buffer layers. The niobium/gold antenna structure (blue diamonds) was deposited without pre-cleaning. The niobium/gold antenna structures were all deposited with pre-cleaning.

layer and gold antenna was above the critical temperature of the detector thin film which results in a single-step drop of the resistance at ≈ 9.5 K which corresponds to the critical temperature of the niobium nitride thin film.

In fact, figure 3.13 shows the superconducting transitions of several different detector/buffer layer combinations which were fabricated using the processes described in this chapter. The temperature of the first drop of the resistance is independent of buffer material and cleaning and is only depending on the thickness of the hot deposited niobium nitride thin film beneath. The buffer layers represented by the blue diamonds and the green triangles are deposited on 5 nm thick niobium nitride films. The buffer layers represented by the red triangles and the black squares were deposited on 6 nm thick niobium nitride films. The influence of the buffer layer is clearly visible in this figure. The detector with the niobium buffer layer deposited without pre-cleaning (blue diamonds) shows a very strong double step transition and a critical temperature of $T_C = 6.8$K. In contrast, a pure niobium film of this thickness has a transition temperature of $T_C = 8.1$K (see figure 3.6). By changing the buffer layer material to niobium nitride and applying pre-cleaning the double step transition can be greatly reduced. For the sample with 20 nm niobium nitride buffer layer the double step transition vanishes completely.

Thus, we can conclude that the process developed and described in this chapter greatly improves the superconducting properties of the detector element. By using a superconducting buffer layer of sufficient thickness the double-step in the superconducting transition can be completely suppressed. The so improved detector shows no more influence of the antenna and contact structures on the critical temperature of the detection element. This in turn leads to a strong reduction of the noise temperature of the detector as will be seen later in this chapter.

3.4.2 Influence of surface pre-cleaning on the critical current of an optimized detector

To analyze the influence of the pre-cleaning and the buffer layer material on the current carrying capabilities of the superconductor the critical current was measured at 4.2 K. The critical current of detectors fabricated from 5 nm thick niobium nitride films with either a niobium buffer layer and no pre-cleaning is $I_{C,Nb} = 180$ µA or a niobium nitride buffer layer with pre-cleaning is $I_{C,NbN} = 190$ µA. The achieved values for the critical current are almost the same for both devices as can be seen in figure 3.14. The 2 K difference in their critical temperature has no influence on the maximum critical current since only the current through the detector element is measured which is identical for both structures. Yet the type of buffer layer and the pre-cleaning procedures have a strong influence on the shape of the current-voltage characteristic (see insert in figure 3.14) close to the critical current. The detector with a niobium nitride buffer layer and a cleaned interface instantly switches from the superconducting into the resistive state at the critical current. In contrast, the device with an niobium buffer layer and an uncleaned interface already shows a very small resistance at currents far below the critical current of the detector thin film.

The insert in figure 3.14 shows that starting at about 25 µA there is a non-zero voltage drop over the detector. This voltage drop increases slowly with increasing current until it switches to the normal conducting state at the critical current of the niobium nitride thin-film detector layer.

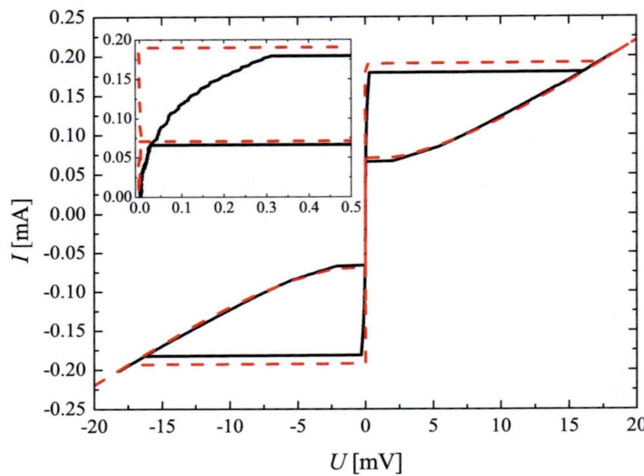

Figure 3.14: *IV* characteristic of a detector with a niobium buffer layer and an uncleaned interface (solid line) and a detector with a niobium nitride buffer layer and a pre-cleaned interface (dashed line). The insert shows a closeup close to I_C. The sample with the uncleaned interface shows a small increasing voltage drop starting from low currents while the pre-cleaned interface shows a clean jump from zero at I_C

The exact origin of this behavior is unclear. One can imagine that due to the regions of suppressed superconductivity below the antenna structure the critical current in these regions is reduced and the superconducting thin film can thus not carry the complete current. A fraction of the current thus flows through the low resistive gold antenna and creates the small voltage drop. A schematic of this model can be seen in figure 3.15 on the left side.

In the regions where the superconductivity is suppressed (shaded orange in figure 3.15) the film already switches to the normal conducting state at a lower critical current $I_{C,\text{suppressed}} < I_2 < I_C$. These regions of suppressed superconductivity are shunted by the above lying low resistive gold antenna. The current that flows through the gold antenna does not produce sufficient heat to switch the complete detector element into the normal conducting state. Thus, only a small voltage drop is observed corresponding to the parts of the current that is shunted through the gold layer. The fraction of the current that is carried by the gold layer and thus the voltage drop increases until the

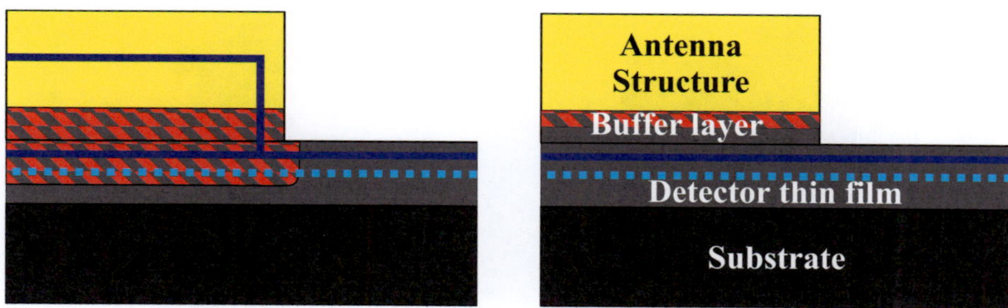

Figure 3.15: Schematic view of the cross section of the detection element with a buffer layer with $T_C < T_{C,\text{detector}}$ on the left side and a buffer layer with $T_C \geq T_{C,\text{detector}}$. The orange shaded regions indicate where the superconductivity of the device is suppressed relative to the detector film. The current flow through the detector at currents $I_1 < I_{C,\text{suppressed}}$ is indicated by the solid dark blue line. Currents with $I_{C,\text{suppressed}} < I_2 < I_{C,\text{detector}}$ are indicated by the dashed light blue line.

critical current of the bridge itself is reached and the complete detector switches into the normal state.

If the buffer layer has a higher critical temperature than the detector film it has no negative influence on the superconductivity in the detector film. The complete current can be carried by the detector film layer. The switching only occurs when the critical current of the detector element is exceeded and it switches into the normal conducting state abruptly.

3.4.3 Characterization of an optimized HEB mixer at 2.5 THz

The HEB mixer detector with optimized interfaces and antenna structure was characterized at a radiation frequency of 2.5 THz. The measurements were performed at the DLR in Berlin by H. Richter. The noise figure of the niobium nitride HEB mixers fabricated using the optimized technologies described in this chapter was measured at 2.5 THz local oscillator frequency. The local oscillator used for these measurements was a quantum cascade laser (QCL) [60]. An image of the complete setup can be seen in figure 3.16 on the left side and a sketch of the interior of the cryo-cooler is shown on the right side. The QCL used as LO is mounted on the 40 K stage of a pulse tube cooler and its output signal is redirected by a lens and a beam splitter back into the

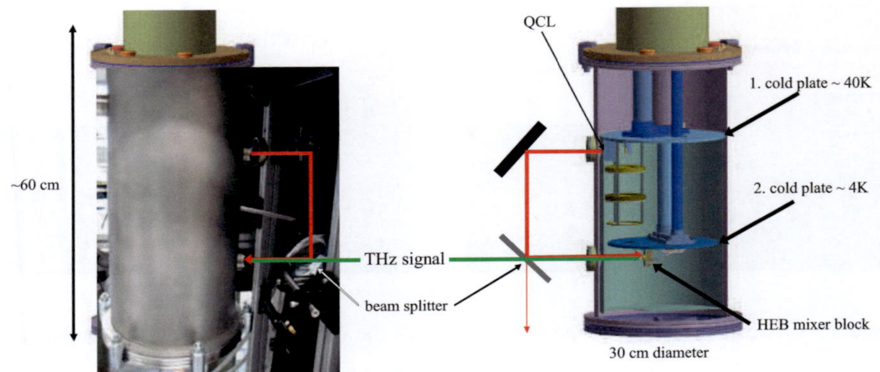

Figure 3.16: Image of the measurement setup at DLR Berlin mounted into the cryo-cooler. A sketch of the interior of the system can be seen on the right side. The QCL that is used as LO is mounted on the 40 K stage of the cryo-cooler and its output signal is redirected by a lens and a beam splitter back into the cryostat. The HEB sample block is mounted on the 4 K stage of the cooler. The THz radiation is then coupled into the HEB through a beam splitter on the same optical path as the QCL power.

cryostat. The HEB sample block is mounted on the 4 K stage of the cooler. The THz radiation is then coupled into the HEB through a beam splitter on the same optical path as the QCL power. The power of the QCL $P_{QCL} = 45\,\mu$W is enough to pump the HEB from the superconducting to the normal conducting state. The pumped IV-curves of the detector can be seen in figure 3.17. The double sideband (DSB) noise temperature was measured using a black body at room temperature as hot load and one cooled to liquid nitrogen temperatures as cold load. The noise temperature was measured using the Y-Factor method explained in chapter 2.5.2. The lowest DSB noise temperature was achieved at a bias current of $I_{bias} = 36\,\mu$A and a bias voltage of $U_{bias} = 0.6$mV. The achieved uncorrected noise temperature was $T_{\text{DSB noise, uncorrected}} = 2000$K. This is the uncorrected value where the losses in the optical path (e.g. beam splitter, silicon lens) and reflection losses are still included in the figure. Correcting this value for all losses in the signal path and the absorption losses of the optical components the DSB noise temperature of the detector can be calculated to $T_{\text{DSB noise,corrected}} = 800$K. This is a reduction of about 25% in noise temperature compared to similar devices fabricated using the same niobium nitride thin films and a 10 nm thick titanium buffer layer.

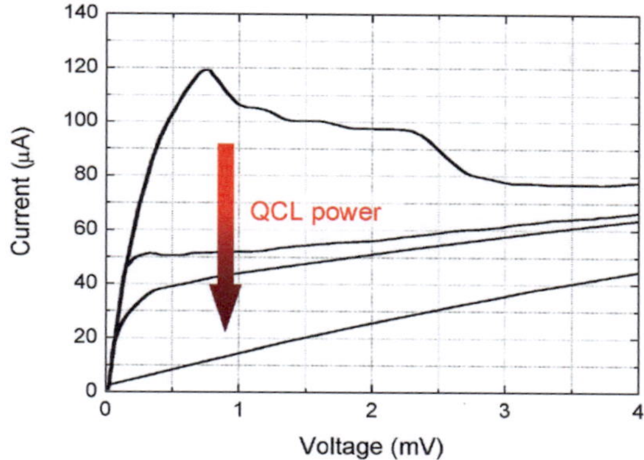

Figure 3.17: *IV*-characteristics of the HEB mixer pumped with a QCL LO at 2.5 THz. The arrow indicates increasing LO power. The black lines correspond to the *IV*-curves pumped with increasing LO power from top to bottom, as indicated by the arrow.

Those devices showed a noise temperature of T_{noise}=1050 K at a radiation frequency of 2.5 THz.

To compare the result of this work with the published data of other groups figure 3.18 shows the DSB noise temperature of several different HEB detectors fabricated by research groups all around the world. The value achieved in this work is marked with a red star. At time of publication in 2009 [60] this was the record noise value obtained for HEB mixers at 2.5 THz. Most recent results were obtained with *in-situ* sputtered gold antennas directly on the niobium nitride ultra-thin film for the detector layer. In [72] an even lower noise temperatures of 600 K at 2.5 THz [72] was claimed. These results further reinforce the assumption that the quality of the interface between the antenna and detector film is one of the main contributors to the noise temperature.

3.5 Conclusion of Chapter 3

We have studied superconducting properties of ultra thin niobium nitride films deposited on different substrates and at different deposition conditions. The superconducting transition temperature of these films shows a strong dependence on the film thickness. It was shown in light of the intrinsic proximity effect theory that this de-

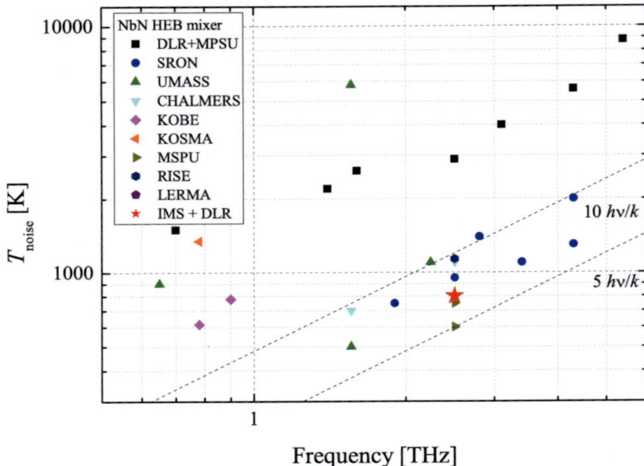

Figure 3.18: DSB noise temperature of several HEB detectors by different groups as indicate by the legend. The DSB noise temperature obtained in this work is marked with a red star. The black line in the figure indicates ten times the quantum limit of the noise. Data was taken from [61, 62, 63, 64, 65, 66, 67, 68, 69, 70, 71, 72].

pendency is due to suppressed superconductivity at the surfaces of the films. We have achieved fabrication of superconducting films deposited on heated silicon substrate with a critical temperature of $T_C \approx 9$ K for 5 nm thick films. This is suitable for fabrication of fast THz HEB detectors.

niobium and niobium nitride thin films deposited at ambient temperatures have been evaluated in respect to their application as superconducting buffer layers between the detector film and a gold antenna. It was found that pre-cleaning of the substrate surface by means of Ar-ion milling greatly improves the normal state and superconducting properties of the films. Reasonable high critical temperature above 9 K were achieved for bi-layers made from both materials. The dependence of the transition temperature and the gold layer thickness was analyzed using the proximity effect theory. From this theory an interface resistance between gold and niobium of $\rho_{int} = 29.8$ and of as low as $\rho_{int} = 17.3$ between gold and niobium nitride was extracted.

Due to their intrinsically higher critical temperature the niobium nitride buffer layers can be made thinner than niobium buffer layers with the same critical temperature and still retain superior superconducting properties. A bi-layer with a critical temperature of above 10 K that is suitable for detector fabrication was developed. This exceeds the transition temperature of the ultra-thin hot-deposited niobium nitride films for the detector element. Such bi-layer are perfect for the usage as antenna layers.

Detectors with a niobium nitride superconducting buffer-layer and pre-cleaning prior to antenna deposition were fabricated. It was shown that such a detector shows a perfect single step resistive transition at T_C. Also their IV-characteristic at 4.2 K have a perfect shape. Noise measurements of this detector showed an uncorrected DSB noise temperature of 2000 K at a radiation frequency of 2.5 THz. After correction for losses the DSB noise temperature of the detector was as low as 800 K. This is a decrease of about 25% when compared to similar devices with a titanium buffer layer. At the time of publication this was the lowest reported noise temperature for a HEB at 2.5 THz. These results are very good when compared with state of the art devices which have a noise temperature between 600 K and 1350 K [70]-[73] at 2.5 THz. This shows that the detectors fabricated in this work are well competitive with state of the art devices.

4 Detector sensitivity modification by thermal coupling engineering

During the last decade the most popular material for development of fast and sensitive detectors in the THz spectral range is niobium nitride (HEB direct detector [74] and mixers [46]). The attractiveness of niobium nitride is caused by its relatively high critical temperature among conventional superconductors, its high hardness, its chemical stability and its developed technology of ultra-thin films on different types of substrates. The thermal coupling G between the thin film used as a detection element and the thermal bath is according to equation (2.2) one of the main influences on the speed of a bolometer and according to equation (2.9) on the other hand it is inversely proportional to the sensitivity of the bolometer. Being able to influence the thermal coupling between film and substrate enables us to change the speed and the sensitivity of the detector accordingly. By this we can tune the detector from being very fast to very sensitive by decreasing the thermal coupling between detector and substrate. Thus, the speed and sensitivity can be adjusted as required by any experiment.

In this chapter, we propose a new approach to influence on the thermal coupling between the detector film and the substrate. On a flat substrate the heat escaping from film to substrate can spread into the whole half-space of the substrate (Figure 4.1). When the substrate at the sides of the superconducting film is removed as depicted on the right side of Figure 4.1 the volume that phonons can escape into is greatly reduced. This reduces the thermal coupling of the superconducting film to the bulk substrate below. In this work a novel model was developed to describe the thermal coupling between superconducting micro-bridges on mesa structures with high aspect ratio with the substrate. In this model we are using one-dimensional thermal balance equations taking into account disordered matter of our thin niobium nitride films and limitations imposed on the phonon mean free path by the width of the silicon mesa. To verify this model the self-heating in long superconducting micro-bridges made from

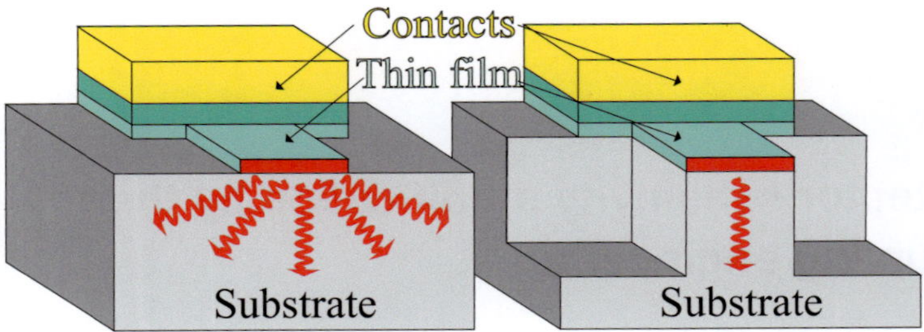

Figure 4.1: On the Left side of this picture the thermal coupling of a thin film micro-bridge to a flat substrate is schematically depicted. In this case the heat (red waves) can escape in the complete half-space which consists of the substrate. On the right side we see the proposed geometry for the restriction of the thermal path. By removing the bulk of the silicon substrate and leaving only a single bridge standing, we reduce the volume that the heat can escape into. In this way we can tune the thermal coupling G between the micro-bridge and the thermal bath.

thin niobium nitride films deposited on top of high silicon mesa structures was studied by analyzing the hysteresis current density j_H (see chapter 4.1). A more than twofold decrease of j_H with the increase in the ratio of heights of the silicon mesa, h, to the width of the micro-bridge, W, from 0 to 24 was observed. Using the developed model a good agreement between theory and experimental data over a wide temperature range from 4.2 K up to the critical temperature T_C for all aspect ratios h/W was obtained. In the last part of the chapter (4.4) we discuss the challenges of fabricating a deep etched detector and show first results of measurement detectors under radiation fabricated with a deep etched detector element.

4.1 Model for the superconducting hysteresis current of deeply etched mesa structures

The hysteresis current is determined by the heat dissipated in the thin film and the thermal coupling of the film to the substrate and thus can be used to analyze the thermal coupling. The hysteresis current I_H is the point where the current in the returning branch of the current-voltage diagram drops back to zero for the first time (see figure 4.2). We expect a constant critical current I_C for structures with different heights h of

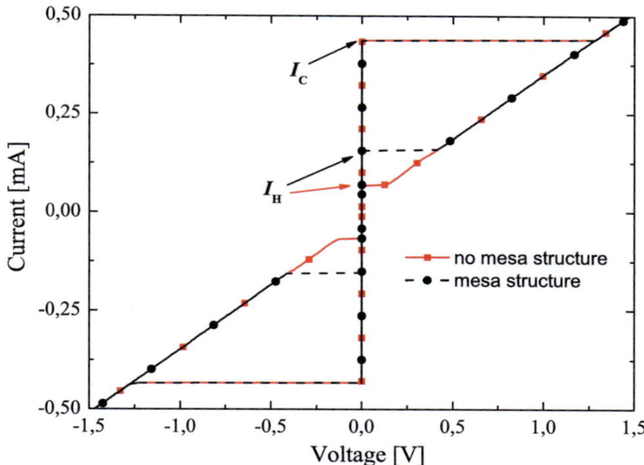

Figure 4.2: Expected shape of the current-voltage diagram of two identical superconducting microbridges on different mesa structures. The curve marked by red squares corresponds to a microbridge on a flat substrate (see left side of figure 4.1). The curve marked by black dots corresponds to a microbridge on a flat substrate (see right side of figure 4.1).

the mesa structure but identical parameters of the microbridge. The hysteresis current in contrast is expected to drop with higher silicon mesas. To be able to compare microbridges with different width we use the hysteresis current density. Using the dimensions of the bridge the hysteresis current density can be calculated to

$$j_\mathrm{H} = \frac{I_\mathrm{H}}{Wd} \tag{4.1}$$

where W is the width and d the thickness of the microbridge. The hysteresis current density should be independent of the width of the film.

The value of the hysteresis current density in thin film superconducting micro-bridges was calculated in numerous works by the analysis of the formation and the dynamics of self-generated hot-spots [35], [75]. These models assume a long superconducting bridge with large metal contacts on a flat substrate. The hot-spot is situated in the center of the bridge and the metal contacts are assumed to be at the same temperature as the thermal bath. The assumption of a long bridge is valid when the bridge is longer than the thermal healing length $\eta = (KdR_\mathrm{th})^{1/2}$. Where K is the thermal conductivity

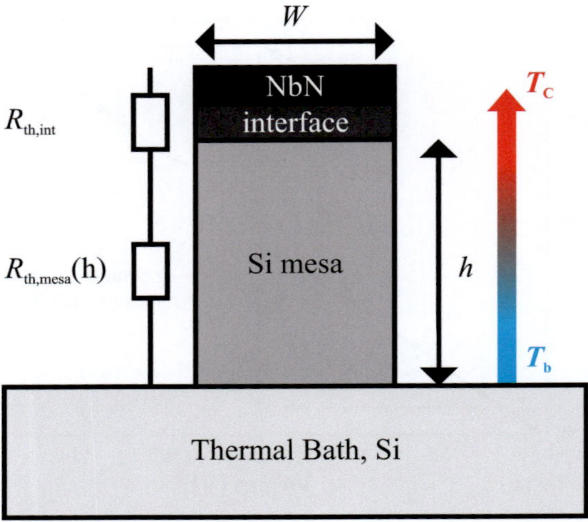

Figure 4.3: Schematic of the cross-section of the silicon mesa structure with the thermal resistances $R_\mathrm{th,int}$ and $R_\mathrm{th,mesa}$ which arise from the interface between film and mesa and the mesa itself respectively. The niobium nitride film is assumed to be at the critical temperature T_C while the bath is assumed to be at bath temperature T_b.

of the film d the film thickness and R_th the thermal resistance between bridge and bath. For the typical structures analyzed in this chapter the healing length is below 100 nm so the model is valid in all cases. In this model the return of a micro-bridge from the resistive to the superconducting state is determined by the thermal equilibrium between generated Joule power I^2R and several cooling channels. The three main of these cooling channels are:

- the cooling through the film-substrate interface, where the substrate is assumed to be semi-infinite and kept at the temperature of the bath;

- the cooling through the surface of the strip, which is performed by radiation and/or thermal contact between the film and cryogenics in gas or liquid form through corresponding interfaces;

- the cooling determined by the thermal conductivity of the metal at a given temperature.

The thermal balance equation describes the film in equilibrium between the heat produced in the micro-bridge and the cooling of the structure through the film and the silicon mesa. Modifying the model given in [35] to match our system the heat balance equation can be written in the following way:

$$I_H^2 R = K_{\text{int}} W L \frac{dT}{dz} + K_{\text{NbN}} W d \frac{dT}{dx} + K_{\text{surface}} W L \frac{dT}{dz} \qquad (4.2)$$

where K_{int} is the thermal conductivity of the interface between thin film and substrate, K_{NbN} is the thermal conductivity of the niobium nitride thin film, K_{surface} is the thermal conductivity of film-to-vacuum/gas/liquid interface. L, W, d are the length, width and thickness of the superconducting microbridge, $\frac{dT}{dz}$ is the temperature gradient in the direction normal to the film surface, i.e. in the film to substrate interface and in film-to-vacuum/gas/liquid interfaces, and $\frac{dT}{dx}$ is the temperature gradient along the length of the thin niobium nitride film bridge.

To model the dependence of the hysteresis current density on aspect ratio for a superconducting thin film on a high mesa structure (see figure 4.3) we followed this approach and modified it according to our particular situation [76]. At first, we can neglect the contribution of the thermal conductivity of the thin niobium nitride layer (the second term in equation (4.2)). Thin niobium nitride films are known as bad normal conductor with a very small electron diffusion coefficient [31] and according to the Wiedemann-Franz law we can estimate the thermal conductivity of thin films from their specific resistivity. For $\rho_{\text{NbN,15nm}} = 4.5 \cdot *10^{-6} \Omega\text{m}$ we can estimate $K_{\text{NbN,15nm}} = 7 \cdot 10^{-2}$ W/(m K) and for $\rho_{\text{NbN,5nm}} = 8 \cdot *10^{-6} \Omega\text{m}$ we get $K_{\text{NbN,5nm}} = 2.8 \cdot 10^{-2}$ W/(m K) at their corresponding critical temperatures. Moreover the cross-section of this channel, Wd, is more than three orders of magnitude smaller than the cross-sections of the other two cooling channels. Thus we can neglect the influence of diffusion cooling. The validity of this assumption will be proven later.

The contribution to the cooling of the bridge through the thin-film surface is significant in case of liquid helium cooling but much smaller in case of helium vapor cooling and performed by radiation only in case of vacuum. At first approximation we thus assume that the contribution of the heat transfer to the surrounding can be neglected. To derive the first term in equation (4.2), which should describe the thermal flow through the film/ substrate interface and further through the silicon mesa, we consider a simple one-dimensional model schematically shown in figure 4.3. In this situation there are two defined temperatures: the temperature of the bridge which equals to the critical temperature T_C at the onset of the superconducting transition and the second is the temperature of the bath T_b measured in the experiment by a temperature sensor. In this case the thermal conductivity through the socket dz/K_{int} can be replaced by the corresponding thermal resistance of the mesa structure and the interface R_{th}. The thermal resistance of the mesa and the interface can be found as

$$R_{\text{th}} = R_{\text{th,int}} + R_{\text{th,mesa}}(h) = R_{\text{th,int}} + \frac{h}{K_{\text{mesa}}} \tag{4.3}$$

where $R_{\text{th,int}}$ is the thermal resistance between thin film and the silicon mesa, and K_{mesa} is the thermal conductivity of the silicon mesa. Inserting equation (4.3) into equation (4.2) we get

$$I_H^2 R = \int_{T_b}^{T_C} \frac{WL}{R_{\text{th,int}} + \frac{h}{K_{\text{mesa}}}} \, dT \tag{4.4}$$

The hysteresis current density can then be written as

$$j_H = \left(\frac{1}{d\rho} \int_{T_b}^{T_C} \left(R_{\text{th,int}} + \frac{h}{K_{\text{mesa}}} \right)^{-1} dT \right)^{\frac{1}{2}} \tag{4.5}$$

where ρ is the resistivity of the film in normal state just above the transition. Using Callaway's model [77] for the lattice thermal conductivity at low temperatures the thermal conductivity can be written as

$$K = \frac{k_B}{2\pi^2 v_{\text{ph}}} \left(\frac{k_B}{\hbar} \right)^3 T^3 \int_0^{\frac{\Theta_D}{T}} \tau(v,T) \frac{x^4 e^x}{(e^x - 1)^2} \, dx \tag{4.6}$$

where v is the phonon frequency, $x = 2\pi\hbar v/k_B T$, where $\hbar$ is the Plank constant, k_B is the Boltzmann constant, Θ_D is the Debye temperature, v_{ph} is the phonon group velocity and $\tau(v,T)$ is the total phonon relaxation time. This model assumes at first a phonon distribution characterized by the Debye spectrum. Furthermore the scattering processes can be described by relaxation times and that the relaxation times can be reciprocally added to a total relaxation time using Matthiessen's Rule:

$$\tau^{-1} = \tau_{\text{ph-b}}^{-1} + \tau_{\text{ph-e}}^{-1} + \tau_{\text{ph-d}}^{-1} + \tau_{\text{ph-ph}}^{-1} \quad , \tag{4.7}$$

where $\tau_{\text{ph-b}}^{-1}$, $\tau_{\text{ph-e}}^{-1}$, $\tau_{\text{ph-d}}^{-1}$, $\tau_{\text{ph-ph}}^{-1}$ are the characteristic times of phonon scattering on boundaries, electrons, defects and other phonons correspondingly. Single-crystal un-doped silicon used for sample fabrication is characterized by a perfect crystalline structure (see for example figure 3.4) and high resistivity even at room temperature ($\rho > 10000\Omega cm$), which increases with decrease in temperature. Therefore we can neglect the contributions to the phonon scattering time τ from the phonon scatter-ing on defects ($\tau_{\text{ph-d}}^{-1}$) and charge carriers ($\tau_{\text{ph-e}}^{-1}$). The contribution through phonon-phonon scattering, $\tau_{\text{ph-ph}}^{-1}$, is also low since the temperature of our measurements ($4.2\text{K} < T < 12.5\text{K}$) is far below the Debye temperature of silicon ($\Theta_{\text{D,Si}} = 645\text{K}$ [9]).

This leaves only boundary scattering $\tau_{\text{ph-b}}^{-1}$ as the main contribution to the thermal conductivity. It has been also shown by Weber et al. [78] that at temperatures below 10K the main contribution to the thermal conductivity K in the bulk silicon arises from the scattering of phonons on surfaces and interfaces. In low dimensional systems being in Casimir limit of very rough surfaces the phonon scattering time $\tau_{\text{ph-b}}^{-1}$ can be estimated as

$$\tau_{\text{ph-b}}^{-1} = \frac{l_{\text{ph}}}{v_{\text{ph}}} \quad , \tag{4.8}$$

where l_{ph} is the mean free path of the phonons. It has been shown by Yong Fu Zhu et al. [79] that the mean free path in thin films can be approximated as

$$l_{\text{ph}} = 2D \tag{4.9}$$

where D is the smallest dimension of the film. In our case the width W of the silicon mesa (which equals to the width of niobium nitride bridge) is the smallest dimen-sion and thus limits the phonon mean free path to $l_{\text{ph}} = 2W$. Therefore the phonon-boundary scattering time (equation (4.8)) is independent of temperature and account-ing for the limit $\Theta_{\text{D}}/T_{\text{C}} >> 1$ the thermal conductivity K (equation (4.6)) can be rewritten as

$$K = \frac{13k_{\text{B}}}{4\pi^5 v_{\text{ph}}^2} \left(\frac{k_{\text{B}}}{\hbar} \right)^3 DT^3 = ADT^3 \tag{4.10}$$

where $D = W$. The thermal resistance is inversely proportional to a thermal conduc-tivity $R_{\text{th}} \propto K^{-1}$. We can thus assume that the temperature dependency of the thermal resistance is $R_{\text{th,int}} = \frac{1}{BT^3}$. Inserting this and equation (4.10) into equation (4.5) for the hysteresis current density we get

$$j_{\text{H}} = \left(\frac{1}{d\rho} \left(\frac{1}{B} + \frac{h}{AW} \right)^{-1} \int_{T_{\text{b}}}^{T_{\text{C}}} T^3 \mathrm{d}T \right)^{\frac{1}{2}} \tag{4.11}$$

For arbitrary bath temperature $T_{\text{b}} < T_{\text{C}}$ the hysteresis current density can be calculated as

$$j_{\text{H}} = \left[\frac{1}{4d\rho} \left(\frac{1}{B} + \frac{h}{AW} \right)^{-1} (T_{\text{C}}^4 - T_{\text{b}}^4) \right]^{\frac{1}{2}} \; . \tag{4.12}$$

With this model the hysteresis current density of long superconducting microbridges on mesa structures with arbitrary aspect ratio can be described.

4.2 Device Fabrication

The thin niobium nitride films, for this study, were deposited by reactive magnetron sputtering of a pure niobium target in argon/nitrogen atmosphere. A more in detail analysis of such niobium nitride thin films is given in chapter 3.1. The films were deposited on $10 \times 10\,\text{mm}^2$ high-resistive ($\rho > 10\ \text{k}\Omega\text{cm}$) silicon substrate heated to about 750°C. The silicon substrate used in this study was undoped silicon, fabricated by the floating-zone method that results in a very low density of defects.

The thickness of the studied niobium nitride films was 15 nm and 5 nm with a critical temperature of $T_{\text{C}} \approx 12.2$ K and $T_{\text{C}} \approx 9.2$ K respectively. The critical temperature was defined as the temperature corresponding to the point on the $R(T)$-curve where the resistance drops to $R \approx 0.5\,\Omega$, defined by the resolution of the measurement system (the zero resistance T_{C}). In a subsequent step the films were patterned into micro-bridges for a four-probe measurement setup using photo-lithography. The width of the ten bridges on one chip varied between 1 and 10 µm. Afterwards the unprotected niobium nitride film was removed by means of argon ion milling down to the silicon surface.

The same resist mask was then used for deep etching of the silicon substrate. The unprotected silicon was etched using a deep reactive ion etching (DRIE, Bosch) process [80]. Deep reactive ion etching was conducted in collaboration with and at the MC2 laboratory of Chalmers University in Gothenburg, Sweden. The working principle

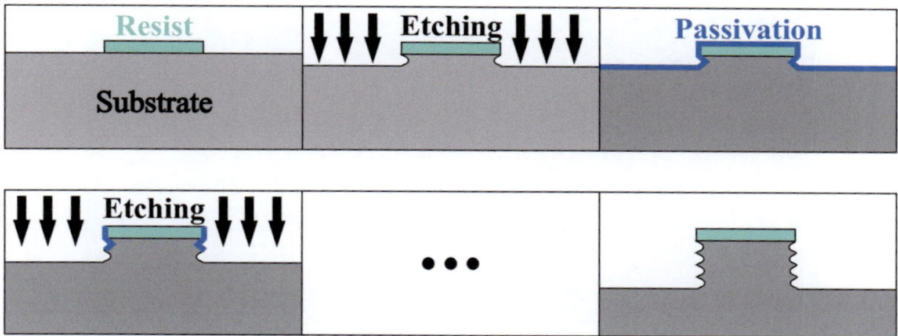

Figure 4.4: The deep reactive ion etching process is a multi-step process depicted above from top left to bottom right. First the structured film is isotropically etched for a few micrometers with a reactive gas. In a second step a passivation layer is deposited on top of the whole sample. Then the passivation layer is removed from all exposed structures parallel to the substrate by physical etching. Subsequently the now exposed substrate is again etched with an isotropic process. Through constant repetition of these steps structures with a very large aspect ratio and perpendicular edges can be fabricated.

of this process is schematically depicted in figure 4.4. This process allows fabrication of deeply etched profiles of silicon with almost vertical edges. To achieve a high anisotropy of the etching the process is split into several steps. By alternating the process gases between SF_6 for etching and C_4F_8 for passivation of the surface these steep edges can be realized. During the SF_6 etching phase the silicon substrate is etched isotropically. In a subsequent C_4F_8 phase a passivation layer of polytetrafluoroethylene is deposited on the whole surface of the to be etched wafer. This layer protects the covered silicon from etching during the isotropic SF_6 etching phase. On surfaces parallel to the substrate surface however, this layer is removed from the silicon during consecutive SF_6 phase by ions accelerated in normal direction to the substrate surface indicated by the black arrows in figure 4.4. This makes the silicon there available for reactive ion etching again. Alternating these steps multiple times a structures with very high aspect ratios can be realized. In the used process a mean silicon etching rate of 2.4 µm per minute was achieved.

With this process it was possible to etch silicon to a depth of about 24 µm and to retain mechanically stable micrometer wide bridges. Deeper etching leads to a punch

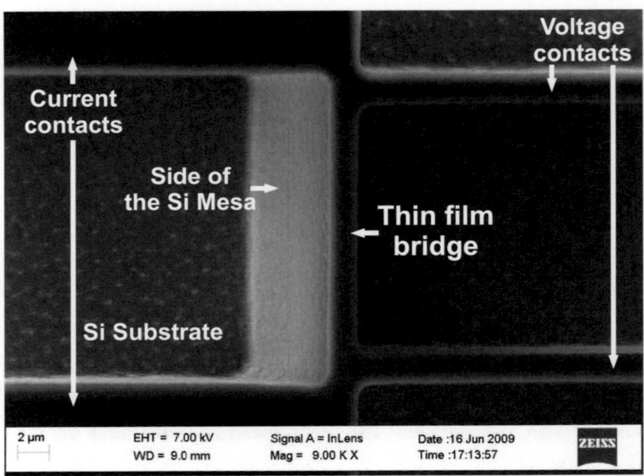

Figure 4.5: SEM image of a 2 µm wide bridge with a length of 20 µm and a height of 24 µm tilted 13° to the right taken at Chalmers MC2 laboratory. In the center the bridge is visible with its contact leads leaving to the top and the right side of the picture. The bright area on the left side of the bridge is the side of the silicon mesa. The lines visible each correspond to one etching step with the DRIE process.

through effect in the smallest 1 µm wide sockets of the micro-bridge which etches away the socket in an uncontrolled manner. Therefore the maximally achievable aspect ratio of etching depth to width of the bridge was about 24, which was obtained for 1 µm wide bridges on a 24 µm high mesa. The actual width W of the micro-bridges was measured with a scanning electron microscope (SEM). The height of the silicon mesa, h, was measured with a Tencor P15 profilometer with a measurement resolution of 1 nm. The SEM image of a typical micro-bridge with a width of 2 µm tilted by 13° to the left is shown in Figure 4.5. The current leads for the four probe measurements are connected at the top and bottom, and the voltage leads are connected via the bridges on the right side. The brighter area on the left side of the bridge is the side of the silicon mesa. The line pattern on the wall is characteristic for samples etched with DRIE and results from the switching between etching and passivation of the surface. Chips with niobium nitride microbridges with different etching depths of the silicon of 0, 8, 16 and 24 µm were fabricated from the 15 nm thick niobium nitride films

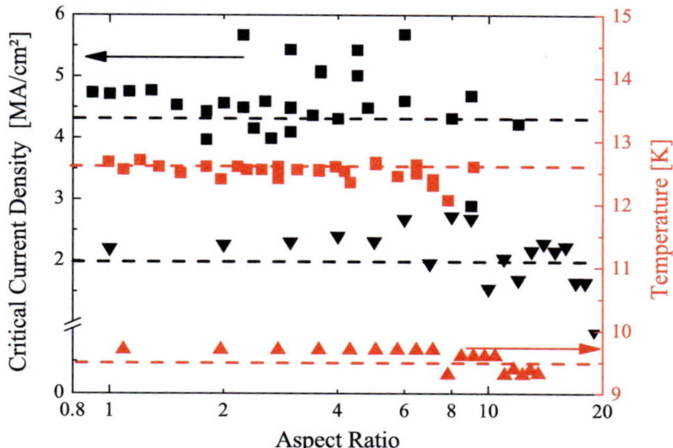

Figure 4.6: Critical current density (black) measured at $T = 4.2K$ in the liquid helium and critical temperature (red) of the niobium nitride microbridges as a function of Aspect Ratio h/W for the 15 nm thick film (Squares) and the 5 nm thick film (triangles) respectively. Both T_C and $j_C(4.2K)$ show only minor fluctuations for structured samples. The dashed lines mark the average value of the respective properties.

and with etching depths of 0, 16 and 24 µm from the 5 nm thick films. Therefore the aspect ratio h/W was varied from 0 (no silicon etching) to 24 (1 µm wide niobium nitride microbridge on 24 µm high silicon mesa).

4.3 Experimental Results

To characterize the superconducting properties of the as deposited films and the afterwards fabricated samples the temperature dependencies of the resistance $R(T)$ and the current-voltage characteristics (IV-curves) were conducted. These measurements of $R(T)$ and IV-curves were performed with a homemade dipstick in a liquid helium transport Dewar. The samples were mounted on a ceramic sample holder allowing simultaneous measurements of up to six samples. The sample holder was placed on a massive copper block in close vicinity to the temperature sensor. Measurements of the $R(T)$-curves were made in a standard four probe experimental set-up to prevent a contribution of the resistance of the bias lines. During the $R(T)$ measurements the sample was biased with a transport current $I_{tr} < 1$ µA which did not overheat the samples even

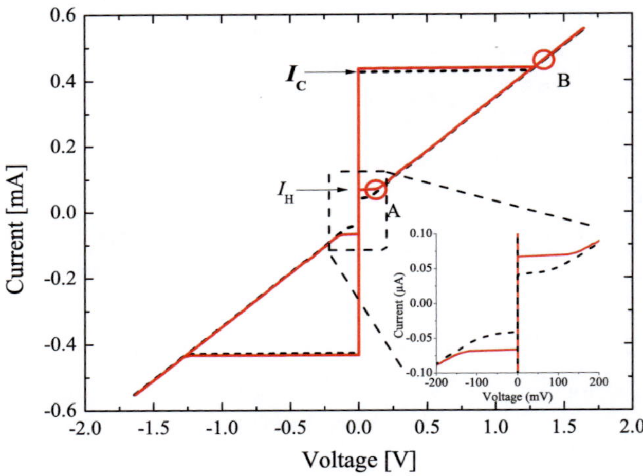

Figure 4.7: Typical current-voltage characteristic of a 15 nm thick niobium nitride film measured over a 1 µm wide bridge in liquid helium (solid line) at 4.2 K and in helium vapor at slightly above 4.2 K (dashed line). Critical current I_C and the hysteresis current I_H are indicated by arrows in the graph. The inset shows a close-up of the hysteresis current. The well visible increase of the hysteresis current is due to an additional cooling channel when immersing the sample into liquid helium. Point A corresponds to the point where the bridge has a temperature of T_C while at point B the bridge is far in the resistive state.

when they had a high normal state resistance. The sample temperature was varied from room temperature to liquid helium temperature using the temperature gradient of the helium gas in the liquid helium dewar. The critical temperature of the finally fabricated micro-bridges does not show any significant change due to degradation of the niobium nitride films that might have been expected from damaging during patterning procedure. Figure 4.6 shows the critical temperature of all studied micro-bridges with different aspect ratios for the 15 nm thick films. Similar results were obtained for the 5 nm thick film.

We obtained a total spread of the critical temperature values smaller than 0.5 K, i.e. less than 5 % of $T_C = 12.2$ K which is identical to the critical temperature of an unpatterned 15 nm thick niobium nitride films on silicon substrate. Measurements of the current-voltage characteristic (*IV*-curves) were made in the current bias mode. Two measured *IV*-curves, one of the sample immersed in liquid helium ($T = 4.2$K) and on of the same sample but kept in helium vapor at temperature only slightly above 4.2K, are shown in figure 4.7. We define the critical current value I_C of our structures as the current at which the micro-bridge switches from the superconducting into the normal state while increasing the transport current. Likewise the hysteresis current I_H is defined as the point in the current-voltage characteristic where the micro-bridge first switches back from the resistive to the superconducting state while decreasing the transport current (see figure 4.2) From the so found values of I_C and I_H the densities of the critical current j_C and the hysteresis current j_H were calculated using the known cross-section of the niobium nitride microbridges, $d \times W$ as $j_{C,H} = \frac{I_{C,H}}{Wd}$. All studied samples showed a strong hysteresis at $T < T_C$. At temperatures very close to T_C ($T \geq 0.98 T_C$) the *IV*-curves became non-hysteretic and therefore the value of the critical current was determined with a 100 µV criterion defined by the resolution of the measurement setup.

When the sample is transferred from the liquid helium to the helium gas phase the hysteresis current of all studied micro-bridges decreased significantly. At the same time the value of the critical current remained almost constant demonstrating only a small reduction as shown in figure 4.7: the I_C value of the 1 µm wide niobium nitride bridges on 9 µm high silicon mesa ($h/W = 9$) decreased from 434 to 427 µA (divergence is less than 2 %) while the I_H value decreased from 67 µA in liquid helium to 41 µA in helium vapor (divergence is almost 40 %).

The values of the critical current density j_C at 4.2 K in liquid helium of all studied niobium nitride microbridges of the 15 nm film and the 5 nm film with different aspect ratios are shown in figure 4.6. In fact, figure 4.6 presents results obtained on several deep etched chips with silicon mesa of different heights between 8 and 24 µm fabricated in separate technological runs. The observed spread of $j_C(4.2\text{K})$, which varies from 4 to 5.6 MA/cm^2, is larger than the spread of T_C of the micro-bridges discussed above. The mechanism of the generation of the critical state in current carrying superconducting bridges is known to be very sensitive to variations in edge quality. Uncontrollable imperfections of the micro-bridge edges made by photo-lithography with the multistep etching process may cause the experimentally observed spread of the $j_C(4.2K)$

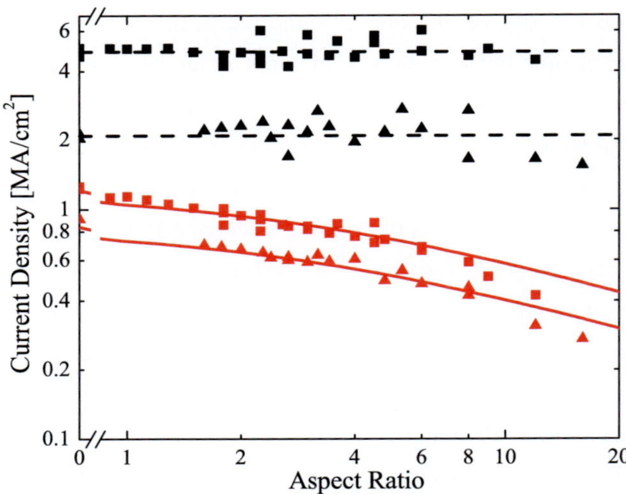

Figure 4.8: Critical current density (black) and hysteresis current density (red) over aspect ratio for a 5 nm (triangle) and 15 nm (square) thick sample respectively at a temperature slightly above 4.2 K (the measurements of IV-curves were made in the helium vapor atmosphere). The solid lines are calculated with (4.12). The dashed lines are to guide the eye.

values. Nevertheless we did not observe any pronounced dependence of the critical current density on aspect ratio as shown in figure 4.6 for both the 5 and the 15 nm film thicknesses. Therefore we conclude that the deep etching process does not influence both the T_C and $j_C(4.2K)$ values and thus that the damage of the micro-bridges caused by this process is minimal. Also the dependence of j_C on the aspect ratio in helium vapor is very similar to what is measured in liquid helium (see figure 4.7). Therefore, the critical current density is insensitive to the presence of liquid helium on the surface of the niobium nitride films in the whole studied range of h/W values.

In contrast to the independence of T_C and j_C from the aspect ratio AR (see figure 4.6) the density of the hysteresis current $j_H(AR)$ (measured in the helium vapor phase at temperatures only slightly above 4.2 K) shows a decrease from $j_{H,15\,nm}(0) = 1.25\frac{MA}{cm^2}$ to $j_{H,15\,nm}(12) = 0.42\frac{MA}{cm^2}$ for the 15 nm thick niobium nitride film and a decrease from $j_{H,5\,nm}(0) = 0.9\frac{MA}{cm^2}$ to $j_{H,5\,nm}(24) = 0.19\frac{MA}{cm^2}$ for the 5 nm thick niobium nitride film respectively. In Figure 4.8 the dependence of j_H on the aspect ratio is shown together with j_C values measured at the same bath temperature $T_b = 4.25$ K and cooling conditions for all samples of the 5 and 15 nm thick niobium nitride films. The solid lines in the plot are calculated using equation (4.12). The theoretical model fits perfectly to the obtained experimental data.

Figure 4.9 shows the dependence of hysteresis current density on reduced temperature T/T_C of two microbridges with different aspect ratio. The solid lines are the fit of the temperature dependency of j_H for two micro-bridges with the different aspect ratios 2.4 and 4 for the 15 nm thick film. For both micro-bridges the experimentally obtained $j_H(T)$ dependencies are described by equation (4.12) in the whole temperature range from 4.2 K up to $T_C = 12.2$ K. Additionally, the dashed line in the same graph shows the temperature dependency of the hysteresis current density obtained with an older model developed by Skocpol et al. in [35]. Using

$$ j_H = \left(\frac{\alpha}{\rho d} \right)^{1/2} (T_C - T_b)^{1/2} \tag{4.13} $$

where α is the heat transfer coefficient per unit area the hysteresis current density is calculated. We notice that for the micro-bridge with $h/W = 2.4$ both calculations coincide at temperatures above $0.9T_C$, while at lower temperatures the curves deviate from each other. At $T = 0.4T_C$ the difference is almost 50 %. The reason for this difference is that in Skocpol et al. [35] it was assumed that the thermal conductivity is independent of the temperature in the observed temperature range. This results in the linear temperature dependency of j_H predicted in equation (4.13). This assumption is correct when the temperature gradient between micro-bridge in the resistive state and bath is much smaller the bath temperature, i.e. considering our situation at T_b slightly lower T_C. For larger differences the temperature dependency of K has to be taken into account which results in the T^4 dependency in equation (4.12). All measured structures show a similar dependency of the hysteresis current on temperature which is independent of aspect ratio or film thickness.

In figure 4.10 the dependency of the temperature gradient $(T_C^4 - T_b^4)$ per sheet dissipated power $(d\rho j_C^2)$ on aspect ratio h/W is shown for all bridges measured at $T = 4.28$K. Independent of the niobium nitride film thickness all experimental points can be described by linear fit as it is predicted by the part of equation (4.12) in parenthesis. The dependency presented in figure 4.10 is temperature independent. This has been

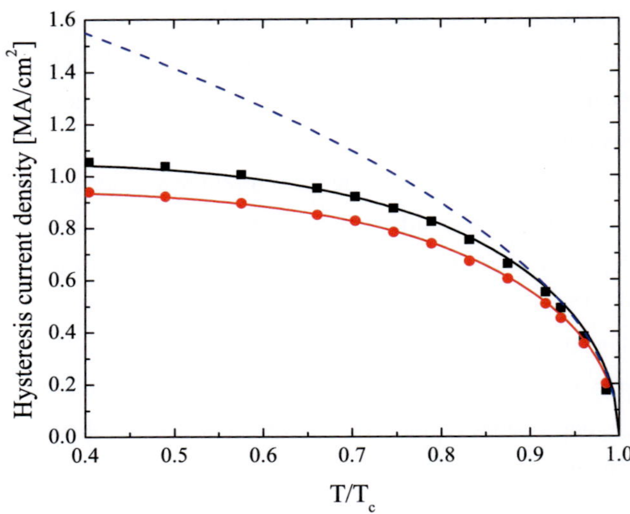

Figure 4.9: Dependence of hysteresis current density on reduced temperature T/T_C of two microbridges with different aspect ratio (black squares ($AR = 2.4$) and red circles ($AR = 4$)). The solid lines are the best fit of experimental results by (4.12). The dashed line is the $j_H(T/T_C)$ dependence taken from (4.13).

proven by the analysis of experimental results on $j_H(h/W)$ performed on the bridges kept at different bath temperatures. From the linear fits shown by the dashed (5 nm thick films) and the solid lines (15 nm thick films) we obtained the values of $A = 4590$ and 4640 W/m^{2}K^4 and $B = 1530$ and 1580 W/m^{2}K^4 for 5 and 15 nm thick niobium nitride films respectively. The A values show almost no difference ($\leq 1\%$). This is expected since A is in fact a material constant determined by the thermal conductivity (see equation (4.6)) of the silicon which is the same for both film thicknesses. Slight differences in B (about 3 %) could be attributed to slightly different growth conditions of the niobium nitride films or the accuracy of the used measurement setup. The obtained parameters A and B are temperature independent in the examined temperature range and thus allow for calculation of the derived physical quantities K_{mesa} and $R_{\mathrm{th,int}}$ at arbitrary temperatures.

Using equation (4.10) we can calculate the thermal conductivity K_{mesa} for the silicon mesa. This yields for a 1 µm wide bridge a thermal conductivity between

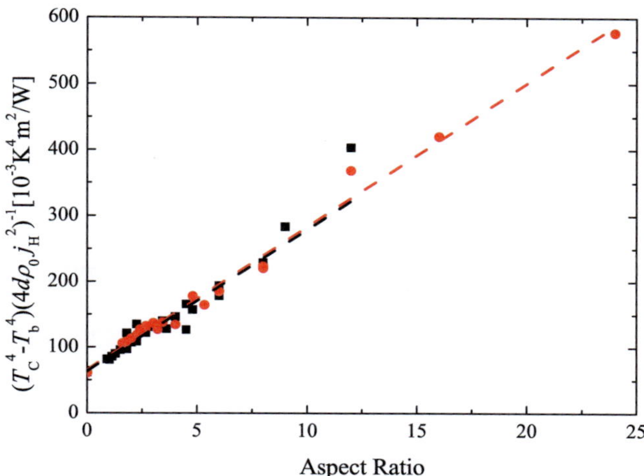

Figure 4.10: Dependency of temperature gradient per sheet dissipated power on the aspect ratio from the data obtained at 4.28 K. The red dots correspond to the 5 nm thick film and the black squares to the 15 nm films respectively. The dashed and solid lines are linear fits for the 5 and 15 nm thick films respectively.

$K_{\mathrm{mesa}}(4.28\,\mathrm{K}) = 0.36\,\mathrm{W/(m*K)}$ and $K_{\mathrm{mesa}}(10\,\mathrm{K}) = 4.6\,\mathrm{W/(m*K)}$. This is well above the value of the thermal conductivity of the niobium nitride thin film estimated at the beginning of this chapter. The thermal resistance $R_{\mathrm{th,int}} = 1/(BT^3)$ calculated for the two different films gives $R_{\mathrm{th,int}}(5\mathrm{nm}, 4.28\mathrm{K}) = 8.3 \cdot 10^{-6}\,(\mathrm{m^2K})/\mathrm{W}$ and $R_{\mathrm{th,int}}(15\mathrm{nm}, 4.28\mathrm{K}) = 8.1 \cdot 10^{-6}\,(\mathrm{m^2K})/\mathrm{W}$. Previous work of Johnson et al. [81] reports values of about $R_{\mathrm{th,int}} = 8 \cdot 10^{-6}\,(\mathrm{m^2K})/\mathrm{W}$ at $T = 4.2$ K comparable to the results of our estimations. Using the results for A and B we can calculate the hysteresis current density using equation (4.12) and results are shown by the solid lines in figure 4.8. The calculated data fits very well with the measured experimental data. This model thus allows us also to predict the hysteresis current density for arbitrary values of aspect ratio and at any bath temperature. The two experimental points in figure 4.10 corresponding to 15 nm thick bridges with $AR = 9$ and 12 deviate from linear fit systematically. Due to imperfect etching conditions a thinning of the silicon mesa in the center of the socket is possible. This can lead to the in chapter 4.2 mentioned punch through effect for the narrowest bridges. By thinning of the silicon socket, the width and thus the thermal escape path of the phonons get restricted stronger than expected from the measured width of bridge which results in a higher thermal resistance. To further increase the thermal resistance and thus reduce the cooling efficiency of thin film superconducting micro bridges one could try to control this effect and construct semi free standing bridges.

4.4 HEB detector for THz radiation with a reduced thermal coupling to the bath

As shown in the previous sections fabricating single microbridges with a high aspect ratio is a controllable process. The fabrication of the micro-bridges is only a single layer process which can be completely done in one photo-lithography step. To fabricate a detector structure multiple fabrication steps are required which are in more detail discussed in chapter 3.4. When fabricating a detector on a silicon mesa one has additionally to take into account the deep etching of the detector element. Therefore the silicon in the surrounding of the detection element has to be removed. Since the width of the detector element is only 1 μm we have to be very careful when etching to prevent the punch through effect mentioned earlier in this Chapter (see section 4.2). Also when etching a micro-bridge structure more than the half of the chip is silicon which is exposed to the reactive gas and the areas to be etched are accessible from all sides. When etching only the center part of the detection element (see figure 4.11) most of the film is covered by resist and the width of the etched area is only a few tens of micrometers which leads to a substantial reduction in the etching rate.

Taking all these challenges into account detector chips with an aspect ratio of 6 (1 μm wide bridge with 6 μm deep etch) were fabricated. In figure 4.11 a SEM image of a detector embedded into a 350 GHz double-slot antenna is shown. In the closeup of the deep etched detection element with the small gap (width $w = 200$ nm $\times$ length $l = 1$ μm) with niobium nitride thin film and the removed substrate in the close vicinity are well visible. The niobium nitride thin film used for this detector has a thickness of $d \approx 5$nm and after fabrication of the detector element a superconducting transition temperature $T_C = 8.9$K (see figure 4.12) which is a standard value for a detector fabricated from such kind of film. The IV- characteristics of the deep etched detector (figure 4.12 (b))

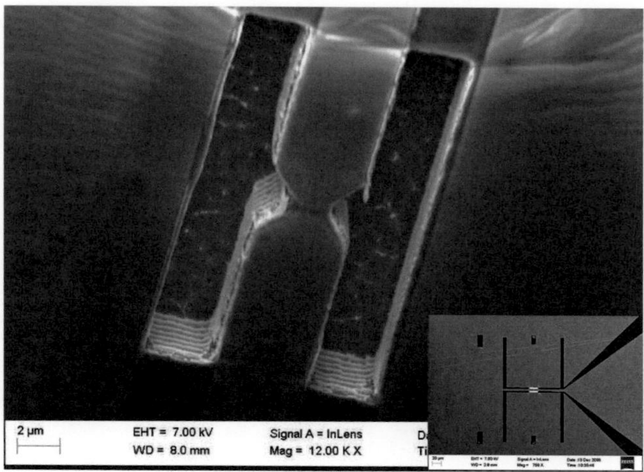

Figure 4.11: This SEM image shows a double-slot antenna detector designed for a frequency of 350 GHz (insert) with a deep etched detector element. The deep etched detector element is rotated and slightly tilted. One can clearly see the sidewalls of the deep etching. The size of the detection element is $w \times l$ 200 nm$\times$ 1 µm

shows a critical current of $I_C = 71$ µA and a hysteresis current $I_H = 37$ µA. For characterization the detector was cooled down in a liquid helium bath cryostat to a bath temperature $T_b = 8.8$K close to T_C. This measurement was conducted several weeks after fabrication of the device and the first characterization of the detector. No degradation of the critical temperature of the device occurred during this period. The operation of the device was tested by pumping the bolometer with a backward wave oscillator (BWO) with a center frequency of 350 GHz as a local oscillator. The output power of the BWO was measured with a Golay cell. A maximum output power of 120 µW was measured. By varying the output power of the local oscillator the detector could be successfully pumped from the superconducting state to the normal conducting state. The $I - V$-curves of the detector with different output power of the local oscillator can be seen in figure 4.13. When increasing the local oscillator output power from zero an increasing suppression of the superconducting transition is observed as expected until the detector switches to a completely normal conducting mode when pumped with very high LO power. The noise temperature of the HEB mixer was measured using a

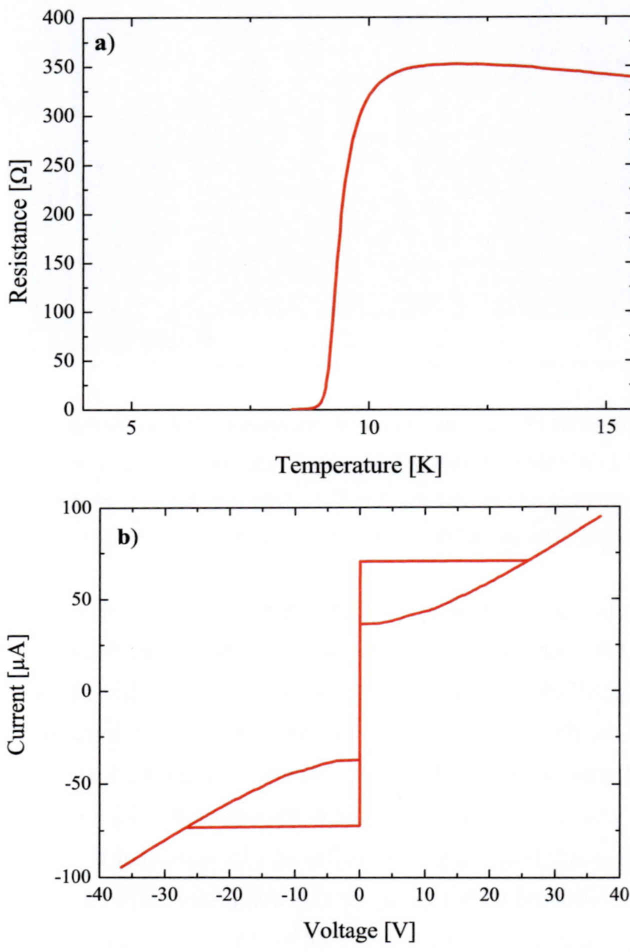

Figure 4.12: Resistance over temperature in the transition region of the deep etched detector element (a). The transition temperature of the deep etched detector is $T_C = 8.9$ K. The IV-curve of the detector immersed in liquid helium shows a critical current of $I_C = 71$ µA and a hysteresis current $I_H = 37$ µA.

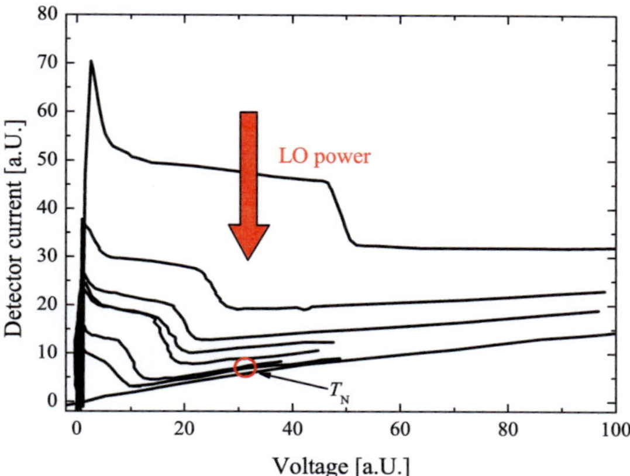

Figure 4.13: Current-voltage characteristics of a deep etched detector pumped at a frequency of 350 GHz. The detector was pumped with a 350 GHz BWO and the lines in the graph are the I-V curves for different output power of the BWO. At the point marked with a red circle and named T_N the lowest noise temperature of the detector has been measured.

black body at room temperature and one cooled down to liquid nitrogen temperatures as hot and cold load respectively. The noise temperature of the detector could then be calculated using the Y-Factor method described in chapter 2.5.2. At the point marked with a red circle in figure 4.13 the lowest noise temperature of the detector with an uncorrected value of $T_N = 10000$ K was measured. At this point the BWO used for pumping had an output power of $P_{BWO} = 31$ μW. This corresponds to an absorbed power in the bolometer of $P_{Bolo} = 22$ nW.

To characterize the direct detection properties of the HEB the detector response to the 350 GHz BWO radiation was measured. In this operation point detector showed a voltage signal of $U_{Bolo} = 204$ mV. Using the output signal of the bolometer and the BWO power absorbed in the bolometer the sensitivity was calculated to $S = \frac{U_{Bolo}}{P_{Bolo}} = 9.3 \cdot 10^6$ V/W. The noise voltage U_{noise} was also measured in this operation point and the measured noise equivalent power (NEP) of the bolometer was $NEP = U_{noise}/S_{Bolo} = 2.4 \cdot 10^{-11} \text{W/Hz}^{1/2}$. The time constant of the detector was evaluated by measuring the IF bandwidth of the superconducting bolometer at 350 GHz to $\tau \approx 300$ ps. This is a strong increase of the time constant compared to typical values for detectors fabricated from 5 nm thick niobium nitride films on a flat substrate [82] which are about 130 ps. By further increasing the aspect ratio of the detector element the increase in time constant might be further increased. This should produce detectors with a higher sensitivity and lower noise due to a larger integration time.

Though the goal of an increased time constant was reached using this method the detector showed a very high noise temperature compared to similar detectors in this frequency range. Standard HEB mixers on flat substrate in this frequency range can show noise temperatures of 600-2000 K (see figure 3.18). An explanation for them high noise temperature might be found in the high operation temperature of the HEB when measuring the noise. According to [83] the minimum noise is reached for $T_{\mathrm{bolo}} \ll T_{\mathrm{C}}$. In our case the detector was operated very close to T_{C}. When operated as a direct detector the HEB showed high sensitivity and a NEP comparable to values obtained for similar niobium nitride direct detectors [84] which showed values between $0.45 \cdot 10^{-11} \mathrm{W/Hz}^{1/2}$ and $1.6 \cdot 10^{-11} \mathrm{W/Hz}^{1/2}$.

The effect of the deep etching on the sensitivity and the *NEP* of the HEB detector was not as large as we expected. The experimental data though shows that a large increase of the time constant of the HEB by deep etching to about 300 ps is possible. This is the basis for further research in this direction.

4.5 Conclusion of Chapter 4

We have studied the superconducting and normal-state properties of niobium nitride micro-bridges with different widths placed on silicon mesa with different heights. Using the Bosch process, structures with a maximum aspect ratio of $h/W \approx 24$ ($h \approx$ 24 μm, $W \approx 1$μm) have been fabricated. No significant degradation of superconducting properties T_{C} and j_{C} of these structures has been observed. We have shown that an increase in h/W results in a decrease of the hysteresis current density by a factor of 3 for the 15 nm thick films with an aspect ratio of $AR = 12$ and a decrease of 4.7 for the 5 nm thick films with an aspect ratio of $AR = 24$. Due to the very small electron diffusivity in thin disordered niobium nitride films and the negligible heat flow from the film to the helium vapor, we consider the heat flow through the silicon mesa as the main cooling channel in our micro-bridges. Solving the one-dimensional heat balance equations we also accounted for the limitation of the phonon mean free path by their scattering on the walls of the silicon mesa. The model obtained under these limiting conditions describes the dependence of the hysteresis current density on aspect ratio h/W well. A linear increase in the thermal resistance was predicted using this model and observed. The experimentally achieved increase in the thermal resistance between the film and the substrate due to deep etching is almost one order of magnitude with respect to the flat silicon substrate.

Fabrication of detector with a deep etched detector element on a silicon mesa with an aspect ratio of 6 succeeded. The characterized detectors showed an increased time constant by a factor of two compared with detectors fabricated on flat silicon substrates. By further increasing the aspect ratio of the structures the speed of the devices might be further reduced which in turn would increase the sensitivity and the decrease the NEP of the HEB. Though the achieved results did not increase the speed by a large value the experimental data suggests that the approach of a deep etched detector element works for increasing time constants of HEB. The developed approach can be used for the adjustment of the cooling efficiency of thin superconducting film microbridges and therefore for tuning the speed and sensitivity of radiation detectors.

5 Measurements of continuous wave and pulsed THz radiation

In this chapter we will show that the detectors developed and fabricated in this work can be readily applied different experimental setups with no change required from the side of the detector. The first part of this chapter contains a brief description of the already available measurement setup in house for characterization of our HEB detectors. The second part of this chapter is focused on the application of the characterized THz detectors. Therefore radiation measurements of two different experiments with different requirements will be shown.

The first experiment was the measurement of fast picosecond THz pulse trains at the ANKA synchrotron. The second experiment was the measurement of the time dependent emission of short pulses and pulse trains by a quantum cascade laser (QCL) to study the thermal properties inside the QCL. This experiment was conducted in collaboration with the university of Leeds. Finally a conclusion on the detector performance and the further outlook for applications will be given.

5.1 Characterization of HEB detectors

The detectors fabricated during this work need to be tested and characterized. For characterization of the HEB detectors at a frequency of 650 GHz an available in-house measurement setup developed in another thesis [39] was used. This system is schematically depicted in figure 5.1.

The system is equipped with a radiation source with a center frequency of 650 GHz and an output power of $P_{rad} \approx 220\,\mu W$. The radiation is coupled out of the source using a horn antenna which results in a linear polarized beam. The beam is focused into the cryostat using metal mirrors.The output power of the THz source can be adjusted by two rotatable wire-grids in the beam path. The focused beam is then radiated into

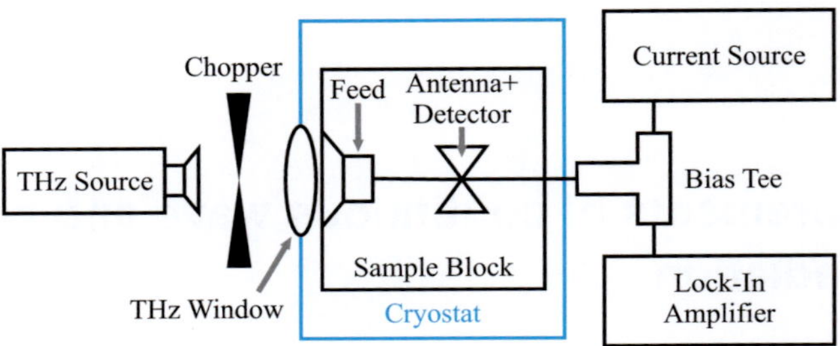

Figure 5.1: Schematic of the measurement setup used to characterize THz detectors [39]. The radiation is provided by a THz source with a frequency of 650 GHz. The radiation from the source is then chopped and coupled via a feed element into the antenna and detector. The detector is read out with either a Lock-In Amplifier or an Oscilloscope. The device is DC biased with a Current source.

a cryostat. The radiation is fed into the cryostat through a THz window fabricated from highly doped PE that is transparent in the THz region. Inside the cryostat the radiation incoming on the sample block is focused on the detector using a silicon lens which is mounted on the backside of the substrate with the detection element. The cryostat used for these measurements is a continuous flow liquid helium cryostat with a minimum reachable temperature of 4.2 K at the cold finger. The cryostat is equipped with several DC lines for biasing of the temperature diodes and heater and a coaxial cable for the readout and biasing of the detector. To measure the temperature at the sample holder the system is equipped with a silicon diode with an accuracy of $T = \pm 0.5$ K. The cold finger is a 1 inch wide copper plate prepared for the mounting of the sample block. To increase the thermal coupling between the sample holder and the cold finger a thin layer of Apiezon N silicon gel is applied on the cold finger as thermal conductor. The detector is biased using a low noise bias box capable of both current and voltage bias operation. The bias box is connected over the RF line to the detector by a room temperature bias tee. The read out of the detectors is done using either a Lock-In amplifier for continuous wave measurements or a fast oscilloscope with a bandwidth of several GHz. In the case that the detector signal needs to further be amplified several different room temperature amplifiers are available in the DC and the RF path.

Figure 5.2: Photo of a sample block from the front side (left) with an already mounted silicon lens. The right side shows a sample block without lens. The detector blocks are made of copper to increase the thermal coupling to the cold finger. The detector chip is embedded into a coplanar waveguide which is used for readout and biasing of the device.

To mount the detector into a cryostat a custom made sample block is needed. It is used as a housing for the detector element and contains a coplanar waveguide with an impedance of 50 Ω into which the detector is embedded. Two different sample blocks used in this work are shown in figure 5.2. On the right side a sample block without additional feed element is depicted. The radiation is directly incident on the detector chip and coupled into the detector element with an planar antenna structure (see chapter 2.4). To reach the low temperatures required for the operation of the HEB detectors the thermal coupling between the cold finger and the detector has to be very good. Therefore the detector block is made from copper which is a material that has a high thermal conductivity even at low temperatures. When the radiation power of the incoming signal is small and/or the beam is not well focused/a planar wave an additional focusing element is required to increase the radiation coupling efficiency. The radiation has to be focused on the antenna after entering the cryostat thus the focusing element has to be on the detector block. On the left side of figure 5.2 the front side of a sample block with a silicon lens as focusing element is shown.

Sensitivity and noise measurements at 650 GHz

Using the measurement setup described in the previous section the sensitivity and noise equivalent power of HEB detectors were measured. Since we had only one available radiation source the detector was characterized in direct detection mode. The characterized detector consists of a 5 nm thick niobium nitride film deposited on sapphire substrate (see chapter 3.1) with a transition temperature of $T_C = 12.2$ K. The antenna is a log spiral antenna fabricated from a 20 nm niobium nitride 200 nm gold bi-layer structure fabricated using the processes described in chapter 3.3. The detector was characterized using continuous wave radiation of the 650 GHz radiation source. For the measurement of the sensitivity the detector was kept at $T_{bath} = 5.7$ K during the experiment which was far below the critical temperature of the detector element. During the experiment the detector was biased with a low noise battery source which was operated in current bias mode. To find the bias point with the highest sensitivity the bias current of the detector I_{bias} was increased from zero up to twice the critical current $I_{C,rad\ off} = 100$ µA and then reduced back to zero. The signal voltage of the detector was measured over the whole range and the resulting data is shown in figure 5.3. While increasing the bias current I_B from zero up to about half the critical current $I_{C,rad\ off}$ no significant detector response is visible. At about half the critical current $I_{C,rad\ off}$ the detector switches into the normal state and shows a small signal response of $U_{sig} = 1.02$ mV. We define this point where the detector switches into the normal state under irradiation as $I_{C,rad\ on} = 50$ µA. When comparing the critical currents we can clearly see a suppression of the critical current $I_{C,rad\ off}$ through the radiation signal. Further increasing the bias current reduces the signal up to zero at a bias current of $I_{bias} = 200$ µA when the detection element has switched fully into the resistive state. When decreasing the bias current the signal rises exponentially until the hysteresis current $I_{H,rad\ on} = 0.3$µA is reached where it shows a peak signal $U_{sig,max} = 3.5$ mV before it suddenly drops to zero when the detector switches back into the superconducting state.

This is the expected behavior of a superconducting detector operated in direct detection mode. The sensitivity of the detector is the highest at the point where the differential resistance of the detector has the highest value (see equation (2.9)). The sensitivity close to $I_{C,rad\ on}$ should also be very high but this is a very unstable point where the

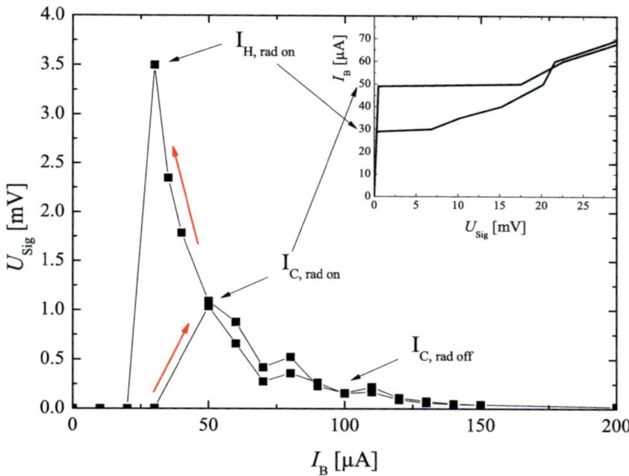

Figure 5.3: Detector voltage over bias current for a typical HEB in direct detection mode. The arrows indicate the critical current $I_{\mathrm{C, rad\ on}}$ and the hysteresis current $I_{\mathrm{H, rad\ on}}$. For comparison the arrow with $I_{\mathrm{C, rad\ off}}$ indicated the critical current without irradiation. The insert shows the current voltage characteristic of the detector under irradiation. The red arrows show the direction the current was changed.

detector cannot be operated due to instantly switching into the resistive state. Taking the point of maximum detector signal and the absorbed power in the bolometer we can calculate the sensitivity of the detector. The system coupling efficiency between the radiation source and the bolometer was evaluated to $\eta_{\mathrm{sys}} = 3.8\%$ by comparing the change of the operation point by absorbed microwave power through the RF line with the change of the operation point induced by incoming radiation. This results in an absorbed power in the bolometer of $P_{\mathrm{abs}} = 8.4$ µW. Using these two values we can calculate the sensitivity of the bolometer according to equation (2.9). This results in a direct detection sensitivity of $S = \frac{U_{\mathrm{sig,max}}}{P_{\mathrm{abs}}} = 416$ V/W. To calculate the noise equivalent power of the detector the noise voltage density was measured with the lock-in amplifier while the radiation incoming on the detector was blocked. The detector was biased during the noise measurement at the same bias point where the highest sensitivity was measured. It was measured to $U_{\mathrm{noise}} = 140\,\frac{\mathrm{nV}}{\sqrt{\mathrm{Hz}}}$. The *NEP* of the detector operated in direct detection mode can be calculated to $NEP = \frac{U_{\mathrm{noise}}}{S} = 3.3 \cdot 10^{-10}\,\frac{\mathrm{W}}{\sqrt{\mathrm{Hz}}}$.

5.2 Measurement of single pulses in electron bunches at ANKA

To determine the escape time of our detector we need to use short THz pulses to excite the detector. From the measured slopes of the pulses we can then calculate the escape time. As a source of fast THz pulses the ANKA synchrotron in Karlsruhe was used.

In a synchrotron electrons are accelerated to relativistic speeds and kept on a trajectory by bending magnets. Every time the electrons are accelerated normal to their trajectory they emit synchrotron radiation [85]. The radiation frequency of the emitted synchrotron radiation is proportional to the length of an electron bunch in the storage ring. When the length of the bunch is chosen shorter than the wavelength of the radiation there is an emission of coherent synchrotron radiation. This radiation in particular in the form of coherent synchrotron radiation is a new kind of source for THz radiation with a high brilliance and large output power. This radiation can be used for several experimental setups like spectroscopy and beam diagnostics of synchrotrons.

When the synchrotron is operated in low-α mode very fast pulses in the THz frequency range are emitted [86]. State of the art THz detectors like silicon bolometers or Schottky diodes have the sensitivity to detect these signals but they have a too long integration time to resolve the single pulses [87]. To measure these fast pulses there is a need for fast superconducting THz detectors.

One kind of superconducting detector suitable for detection of THz radiation in a synchrotron is the niobium nitride hot-electron bolometer [88]. The response times of a few hundreds of picoseconds is fast enough to resolve single pulses. The detector described in chapter 5.1 was used to measure the radiation output of the ANKA synchrotron in the low-α mode. The system used for measuring the radiation is similar to the one described in chapter 5.1 except that the signal was measured with a fast oscilloscope with a bandwidth of 26 GHz. The detector was biased over a bias-tee and the same continuous flow cryostat was used for the cooling of the detection element.

In figure 5.4 a pulse train emitted from the synchrotron is shown. The relaxation level of the detector when there is no radiation incoming is at about -20 mV. When radiation is incoming on the detector the level to witch the detector is relaxing lies at about 0 mV. This is probably due to a beginning of an over-saturation of the THz detectors. This might happen since the output power of the synchrotron is large and the niobium

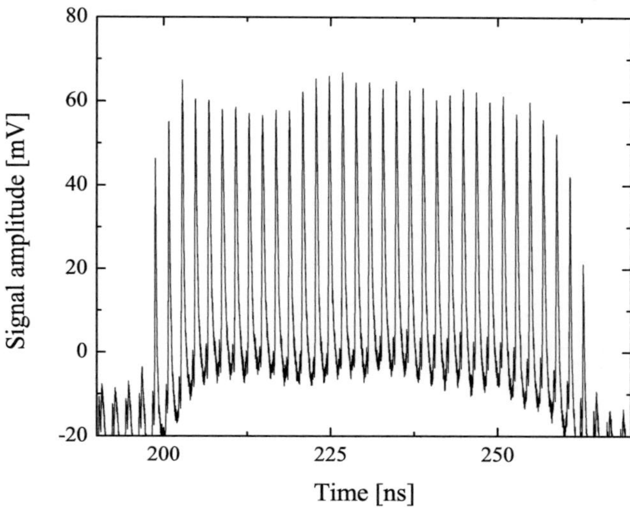

Figure 5.4: Measurement of a pulse train emitted by the ANKA synchrotron. The pulses were directly measured using the niobium nitride HEB described in this chapter.

nitride HEB has a limited dynamic range. When zooming in to the pulses of the pulse train the single pulses can be clearly distinguished. Figure 5.5 shows in a) the repetition rate of the single pulses which is exactly $\tau_{\text{rep}} = 2$ ns. This corresponds perfectly to the 500 MHz frequency of the RF system used to pump the synchrotron.

In figure 5.5 b) the time scale is further increased to show a single pulse at the beginning of a pulse train. The exponential decay of the HEB signal on the falling edge is nicely visible. The decay time of the HEB of $\tau_{\text{pulse}} = 185$ ps was obtained by exponentially fitting the falling edge of the pulse. The obtained decay time is comparable to the bolometric response time [82] of similar detectors. Using the decay time of the detector in direct detection mode an estimate of the possible IF bandwidth in mixing can be made. The estimated IF bandwidth for such a detector can be calculated to $f_{\text{IF}} = \frac{2\pi}{\tau_{\text{pulse}}} \approx 3.4$ GHz. This is comparable to results obtained by various different groups where th IF bandwidth ranges from 2 to 4 GHz [89]. Higher IF bandwidth can be realized when large noise temperatures are acceptable.

When measuring the time constants of pulses further inside the pulse train an increase of the time constant up to $\tau_{\text{pulse,max}} = 430$ ps for pulses on the top of the plateau can be measured. This is a further indicator for an overheating of the bolometer with the incoming synchrotron radiation and a switch to a classical bolometric operation mode. To prevent overheating of the detectors during the operation at the synchrotron the output power of the synchrotron has to be attenuated. Then the decay time of the single pulses might decrease slightly due to no longer being overheated. To further increase the time resolution of the pulse measurements the bolometers need to be tuned to operate as close at their intrinsic time constants as possible. The limiting time constant of the bolometer is according to chapter 2.1.2 the escape time of the thermal energy into the substrate. By further reducing the film thickness or the volume of the detector element the speed could be improved. But fabricating detectors with film thicknesses much lower than 5 nm while still retaining good superconducting properties is a great challenge.

5.3 Analysis of quantum cascade laser radiation characteristics with THz bolometers

Quantum cascade lasers (QCL) are a novel type of radiation source [90] which is able to emit in the THz and infrared frequency range. In contrast to the classical semiconductor lasers, where the emission frequency is defined by the band gap of the semiconductor, the emission of a quantum cascade laser is defined by the semiconductor heterostructure. Being determined by the design of the QCL the output frequency of such lasers can cover a very wide frequency range from the mid-infrared to the submillimeter wave region. At the same time being solid state lasers they are very compact and are thus suitable for integration in mixer experiments (see figure 3.16). This makes them very interesting as local oscillators in the THz frequency range. Another advantage of the QCL is that they can be electrically tuned [91] which allows for a larger spectrum to be covered only by a single laser.

Since quantum cascade lasers in the THz region are quite novel devices there is still a lot of ongoing research about the internal processes inside the laser and its emission characteristics. Since the output power of the quantum cascade laser is limited [92], the detectors to characterize the devices have to be very sensitive. The standard Golay-Cells or Germanium bolometers which are normally used to characterize

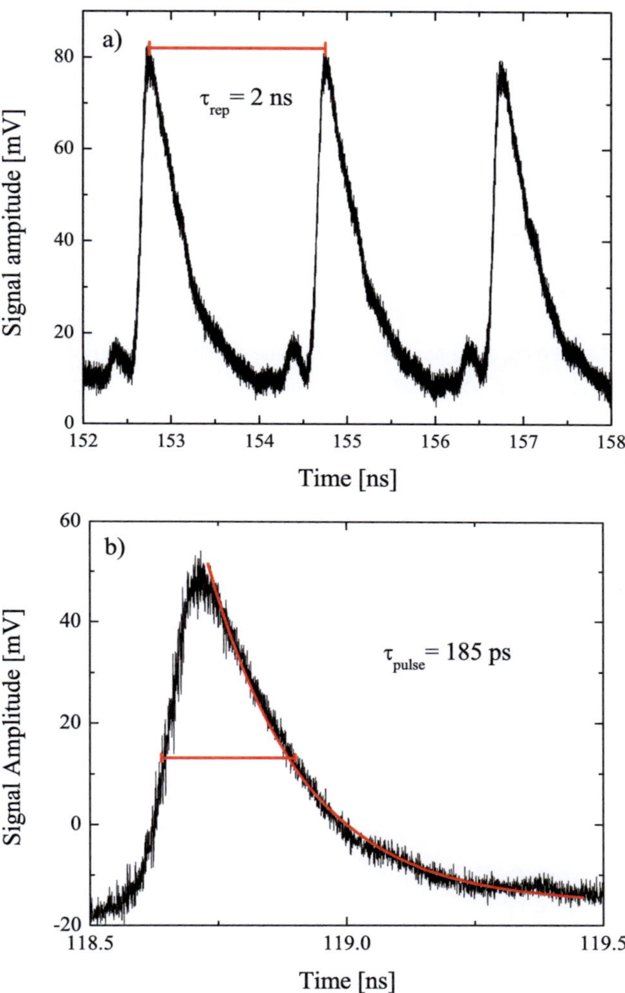

Figure 5.5: Zoom in on the detector measurement of three pulses in the center of a pulse train (a). The repetition rate of the pulses in a train is $\tau_{rep} = 2$ ns. In (b) a closeup on a single pulse is shown. The decay time of the pulse duration is $\tau_{pulse} = 185$ ps.

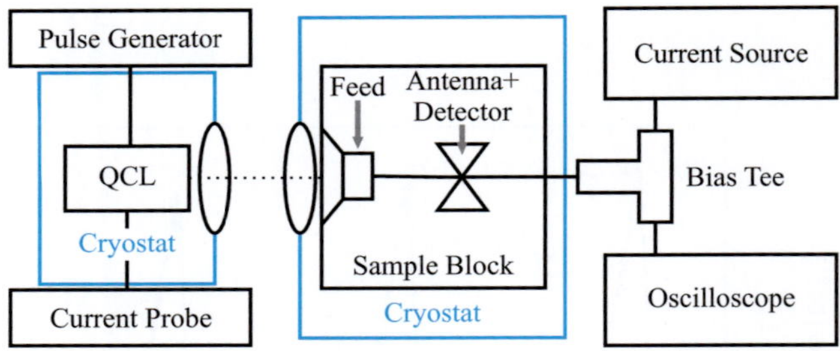

Figure 5.6: A schematic of the measurement setup used to characterize the QCL. The radiation is provided by a the QCL which is supplied by a pulse generator. The current flowing through the QCL is measured by a current probe. The radiation from the QCL is coupled into the antenna and detector. The detector is read out with an Oscilloscope. The device is DC biased with a Current source.[93]

the lasers have the required sensitivity but with time constants in the range of microseconds lack the speed to resolve fast QCL pulses with a duration of a few hundred nanoseconds. Therefore only the average output power of the laser can be measured using these detectors. In contrast using the fast and sensitive detectors fabricated in this work we expect to measure the time dependent emission of the QCL and not a time averaged value. Knowledge about this time dependent emission is of great interest for QCL researchers to better understand and improve the performance of QCL devices.

To measure the fast features of the QCL pulses a fast niobium nitride HEB developed in this work was used. The measurement setup used for characterization of the QCL is depicted in Figure 5.6. The measurements were conducted by Alexander Scheuring in cooperation with the University of Leeds [93].

For the measurements the QCL was operated in a continuous flow liquid helium cryostat and was kept at a temperature of $T_{\mathrm{QCL}} = 15$ K. The laser was biased with a pulse generator which delivered pulses with a set voltage level, and the current flowing through the QCL was measured using a current probe. The output frequency of the QCL was at about 3 THz. The bolometer was mounted in a different continuous flow cryostat opposing the cryostat of the QCL and was kept at $T_{\mathrm{HEB}} = 11$ K during the experiment. The output signal of the bolometer was measured with a 500 MHz real-time oscilloscope. The bolometer was biased using a low noise source operated in voltage bias mode.

For the experiment the pulse generator applies a set voltage to the QCL for 500 ns with a repetition rate of 10 kHz. The current flowing through the QCL during this time stimulates the emission of THz radiation. To be able to monitor the current flowing through the QCL it is measured using a current probe. The detector response to the emitted THz radiation is measured using the oscilloscope. During the experiment the QCL was biased with voltage pulses between 12.0 V and 16.6 V. Figure 5.7 a) shows a sample of four different driving currents for different QCL voltage pulses between $U_b = 12.6$ V and $U_b = 14.1$ V. When increasing the Voltage at the pulse generator the shape of the current flowing through the QCL stays the same. For higher voltages set the absolute values of the current flowing through the QCL increase. It can be seen that the current shows some strong fluctuations in the first 150 ns of the pulse and then smoothly rises during the pulse duration. At the beginning and the end of the pulse small peaks in the current can be seen probably originating from the switching of the pulse generator. In Figure 5.7 b) the voltage response of the detector corresponding to the power emitted during a pulse can be seen. Since the output power of the QCL is not sufficient to overheat the HEB the detector is working in the linear regime. When operating in the linear regime the emission of the QCL linearly corresponds to the voltage output of the detector. The emission of the laser seems to be strongly fluctuating in the first 150 ns of the pulses. This corresponds to the before mentioned oscillations of the laser current at the beginning of the driving pulse. In the second part of the pulse where the current remains almost constant the emission is also very stable. From this we can see that our detector is suitable to follow the time dependent emission of the QCL. Measurements with e.g. niobium air-bridge bolometers [94] didn't allow for this exact point by point mapping of the driving current due to a too large time constant.

Using these result it is possible to predict the behavior of the QCL to a driving current of any arbitrary form and the time dependent output power of the laser pulse can be simulated. Those simulations show a very good agreement with the measured pulses. The response of our detector was fast enough to resolve even small features of the QCL emission. From the emission data it was possible to calculate the power transfer function and exactly evaluate the threshold and cutoff currents of the QCL [93]. These

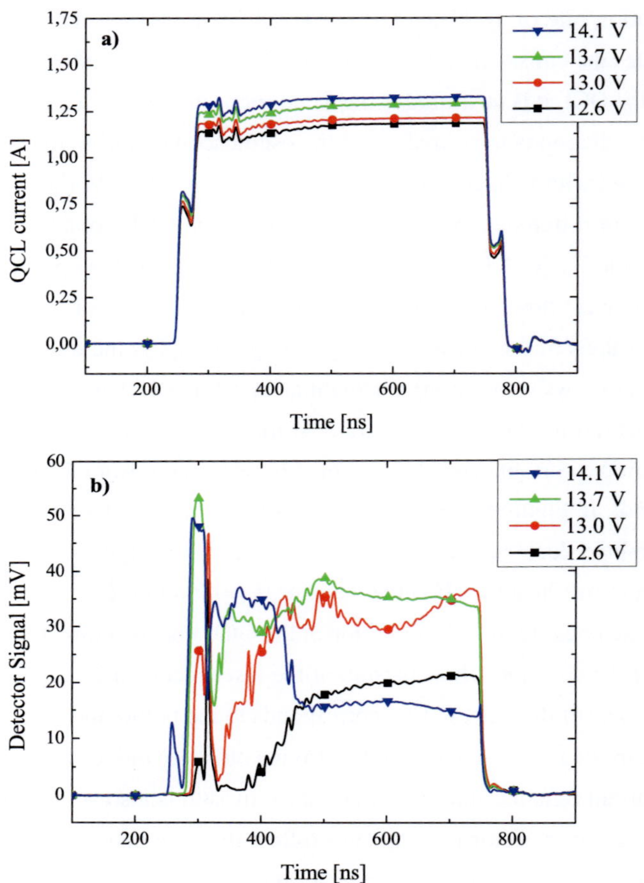

Figure 5.7: Picture a) shows the driving current of the QCL for different voltage levels ranging from $U_b = 12.6$ V to $U_b = 14.1$ V. In b) the corresponding emission of the QCL as measured by the niobium nitride HEB detector is shown. The fast features of the QCL driving current are very well visible in the detector measurements.

measurements allow a simulation of the THz pulse forms emitted by the QCL when driven by a certain current pulse.

5.4 Conclusion of Chapter 5

Hot-electron bolometer detectors for experimental applications were used as direct detectors in several different experimental setups. From the obtained measurement data the figures of merit for the direct detection mode were extracted. The sensitivity of the bolometer was measured at 650 GHz. The maximum sensitivity of the detector in the optimal operation point was $S = 416$ V/W This is considerably smaller than the values reported in the previous chapter but sufficient for all experimental applications. In the same setup the noise of the bolometer was measured and the noise equivalent power of the bolometer was evaluated to $NEP = 3.3 \cdot 10^{-10} \frac{W}{\sqrt{Hz}}$. From the measurements of fast THz pulses the time constant of the bolometer was evaluated to be as low as $\tau_{\text{dec}} = 185$ ps. This corresponds roughly to a maximum frequency of $f_{\text{IF}} = 3.4$ GHz which should be the maximum achievable intermediate frequency for this detector when operation as a HEB mixer.

The detectors fabricated in this work were applied in two distinct setups.

- Using these detectors the coherent synchrotron radiation emitted by the ANKA synchrotron was measured. The sensitivity and speed of the bolometer was sufficient to resolve single electron bunches in the synchrotron pulse trains. It was also possible to determine the response time of the bolometer to 185 ps. This corresponds to an estimated IF bandwidth of 3.4 GHz for a detector operated as a HEB mixer.

- Using the detector to characterize the emission of a quantum cascade laser it was for the first time possible to measure the time dependent output of these QCL devices and resolve the fast features of the QCL output. From the time dependent output the power transfer function could be calculated. From this the threshold $I_{\text{threshold}}$ and cut-off current $I_{\text{cut-off}}$ of the QCL could be evaluated.

Detectors fabricated in this work have been shown to be robust and easily adaptable to different kinds of experimental setups. They show high enough sensitivity and a low *NEP* to be applied in actual experimental setups outside of controlled laboratory conditions. It was shown that the detectors are fast enough to resolve features in fast radiation sources like synchrotron radiation or QCL pulses.

6 Summary

The aim of this work was the optimization of hot-electron bolometers fabricated from ultra-thin niobium nitride films. Therefore the interfaces between the different parts of the hot-electron bolometer, namely the antenna structure the bolometer thin film and the substrate were analyzed. Also the fabrication processes for these structures were optimized in light of the application for HEB devices. Basis for the optimization was the already existing design of hot-electron bolometers fabricated from niobium nitride thin films on silicon. These detectors are embedded in a gold antenna designed for the THz frequency range.

To improve upon the detector performance in a first step the deposition conditions of the superconducting detector film were optimized. Therefor the superconducting properties of the thin niobium nitride films deposited on different substrates at different deposition conditions were analyzed. The superconducting transition of these thin films shows a strong dependence on the film thickness the substrate choice and the deposition temperature. An improved superconducting transition temperature was achieved when depositing the films at high temperatures which could be attributed to better crystalline growth. The strong thickness dependence of the transition temperature could be explained using the intrinsic proximity effect theory. The fabrication of ultra-thin superconducting niobium nitride films on silicon substrate with a thickness of 5 nm and a transition temperature of $T_C \approx 9$ K was achieved in this work. These films are well suited as the basic material for the fabrication of HEB detectors.

For improved adhesion and improved electrical and superconducting stability an *in-situ* superconducting buffer layer was developed to be deposited between the gold and the niobium nitride layer. Therefor the superconducting properties of niobium and niobium nitride thin films deposited at ambient temperatures were analyzed in respect to their application as superconducting buffer layers. It was found that pre-cleaning the substrate surface by means of Ar-Ion milling greatly improves upon the normal state and superconducting properties of the films. Both materials showed promising results for application as buffer layer but due to the intrinsically higher T_C buffer layers made from niobium nitride seem more promising.

In a second step the bi-layers of niobium/gold and niobium nitride/gold as used for the antenna structure were analyzed. From the superconducting transition temperatures the interface resistance between the gold and the superconducting film was analyzed using the proximity effect theory. An interface resistance between gold and niobium of $\rho_{int} = 29.8$ and of as low as $\rho_{int} = 17.3$ between gold and niobium nitride was extracted. From these results the niobium nitride/gold bi-layers were evaluated as the optimal choice for antenna fabrication.

Detectors with an *in-situ* niobium nitride buffer-layer between the antenna and the bolometer film were fabricated. Prior to the antenna bi-layer deposition the bolometer film was pre-cleaned . It was shown that such a detector shows a perfect superconducting transition at I_C in its IV-characteristic at 4.2 K. Noise measurements of this detector showed an uncorrected DSB noise temperature of 2000 K at a radiation frequency of 2.5 THz. After correction for losses the DSB noise temperature of the detector was as low as 800 K. At time of publication in 2009 [60] this was a record noise value obtained for HEB mixers at 2.5 THz. Also this value is an improvement of about 25% compared to similar devices with a titanium buffer layer.

The thermal coupling between the thin film and the substrate was analyzed as means to increase the sensitivity of the HEB detector. To thermally decouple the superconducting niobium nitride film from the substrate the complete superconducting bridge was placed on a silicon mesa. These silicon mesas were structured into silicon substrate using a deep reactive ion etching process. To analyze the influence of these mesa structure on the thermal coupling the superconducting and normal state properties of the niobium nitride bridges were measured and analyzed. The ratio of height to width of the mesa, the aspect ratio h/W was varied from 0 up to 24. No degradation of the superconducting properties T_C and j_C was observed due to the fabrication process. The value of the hysteresis current density though was shown to increase by a factor of 3 for 15 nm thick niobium nitride films with an aspect ratio of $AR = 12$ and 4.7 for 5 nm thick films with an aspect ratio of $AR = 24$. To explain this behavior a model was developed that considered the heat flow through the mesa as the main cooling channel for the microbridge. Solving one-dimensional heat balance equations and accounting for the limitation of the phonon mean free path by surface scattering on the walls of the silicon mesa this model describes the obtained experimental results very well. A linear increase in the thermal resistance was predicted using this model and also experimentally observed. The achieved increase of the thermal resistance is almost one order of magnitude when compared with films deposited on flat silicon substrate.

Detectors with a deep etched detector element were fabricated and showed an increased time constant by a factor of two at an aspect ratio of 6. The sensitivity and noise equivalent power of these detectors were measured to $S = 9.3 \cdot 10^6$ V/W and $NEP = 2.4 \cdot 10^{-11}$ W/Hz$^{1/2}$. By further increasing the aspect ratio of such detectors an increase in sensitivity and a decrease of the NEP is expected. The developed approach can be used for the adjustment of the cooling efficiency of thin superconducting film micro-bridges and therefore for tuning the speed and sensitivity of radiation detectors. The detectors fabricated and developed in this work were also deployed to different experiments. The relaxation time of the detector was measured to be as low as $\tau_{\mathrm{dec}} = 185$ ps using coherent synchrotron radiation. This corresponds roughly to a maximum frequency of $f_{\mathrm{IF}} = 3.4$ GHz. This frequency should also be the maximum achievable intermediate frequency for this detector when operation as a HEB mixer.

The detectors were used to characterize the coherent synchrotron radiation emitted by the ANKA synchrotron. The sensitivity and speed of the bolometer was sufficient to resolve single electron bunches in the synchrotron pulse trains. The detectors were also used to characterize the emission of a quantum cascade laser. It was for the first time possible to measure the time dependent output power of these QCL devices. From the time dependent output power the power transfer function of the QCL could be measured.

Detectors, fabricated in this work, have been shown to be robust and easily adaptable to different kinds of experimental requirements. They show high enough sensitivity and a low NEP to be applied in actual experimental setups outside of controlled laboratory conditions. It was shown that the detectors are fast enough to resolve features in fast radiation sources like synchrotron radiation or QCL pulses.

List of Figures

List of Tables

List of Own Publications

Publications in peer-reviewed journals

K.S. Il'in, A. Stockhausen, A. Scheuring, M. Siegel, A.D. Semenov, H. Richter, and H.-W. Huebers. Technology and performance of thz hot-electron bolometer mixers. *Applied Superconductivity, IEEE Transactions on*, 19(3):269-273, June 2009. doi: 10.1109/TASC.2009.2018266

A. Stockhausen, K. Il'in, M. Siegel, U. Södervall, P. Jedrasik, A. Semenov, H.-W. Hübers. Adjustment of self-heating in long superconducting thin film NbN micro-bridges
Superconductor Science and Technology 25(3):035012, 2012.
doi: 10.1088/0953-2048/25/3/035012

A. Scheuring, P. Dean, A. Valavanis, A. Stockhausen, P. Thoma, M. Salih, S. Khanna, S. Chowdhury, J. Cooper, A. Grier, S. Wuensch, K. Il'in, E. Linfield, A. Davies, and M. Siegel. Transient Analysis of THz-QCL Pulses Using NbN and YBCO Superconducting Detectors. *Terahertz Science and Technology, IEEE Transactions on*, 3(2):172-179, 2013.
doi: 10.1109/TTHZ.2012.2228368

A. Valavanis, P. Dean, A. Scheuring, M. Salih, A. Stockhausen, S. Wuensch, K. Il'in, S. Chowdhury, S. P. Khanna, M. Siegel, A. G. Davies, and E. H. Linfield. Time-resolved measurement of pulse-to-pulse heating effects in a terahertz quantum cascade laser using an NbN superconducting detector. *submitted for publication*

Publications in non peer-reviewed journals

P. Dean, A. Valavanis, A. Scheuring, A. Stockhausen, P. Probst, M. Salih, S. P. Khanna, S. Chowdhury, S. Wuensch, K. Il'in, E. H. Linfield, A. G. Davies and M. Siegel. Ultrafast sampling of terahertz pulses from a quantum cascade laser using superconducting antenna-coupled NbN and YBCO detectors. *Proceedings of the 37th International Conference on Infrared, Millimeter, and Terahertz Waves (IRMMW-THz)* 2012
doi: 10.1109/IRMMW-THz.2012.6380341

A. Scheuring, P. Probst, A. Stockhausen, K. Ilin, M. Siegel, T. A. Scherer, A. Meier and D. Strauss. Dielectric RF properties of CVD diamond disks from sub-mm wave to THz frequencies. *Proceedings of the 35th International Conference on Infrared, Millimeter and THz Waves (IRMMW-THz)* 2010
doi: 10.1109/ICIMW.2010.5612543

A. Scheuring, A. Stockhausen, S. Wuensch, K. Il'in and M. Siegel. A new analytical model for log-periodic Terahertz antennas *Proceedings of the Fourth European Conference on Antennas and Propagation (EuCAP)* 1-5, 12-16 April 2010.

K. S. Il'in, A. Stockhausen, M. Siegel, A.D. Semenov, H. Richter and H.-W. Hübers. NbN HEB for THz Radiation: Technological Issues and Proximity Effect. *Proceedings of the Ninteenth International Symposium on Space Terahertz Technology* April 2008

Supervised Student Theses

"Herstellung und Charakterisierung dünner Detektorfilme für THz-Sensoren", Studienarbeit, Institut für Mikro- und Nanoelektronische Systeme, Karlsruher Institut für Technologie (KIT) 2010, Muhammet Albayrak

"Untersuchung von Isolatorschichten zur thermischen Entkopplung von Bolometern", Institut für Mikro- und Nanoelektronische Systeme, Karlsruher Institut für Technologie (KIT) 2010, Sandro-Diego Wölfle

Bibliography

[1] F. Sizov. THz radiation sensors. *Opto-Electronics Review*, 18:10–36, 2010. doi: 10.2478/s11772-009-0029-4.

[2] C. Lodewijk, T. Zijlstra, S. Zhu, F. Mena, A. Baryshev, and T. Klapwijk. Bandwidth Limitations of Nb/AlN/Nb SIS Mixers Around 700 GHz. *Applied Superconductivity, IEEE Transactions on*, 19(3):395–399, june 2009. doi: 10.1109/TASC.2009.2018231.

[3] M. J. Feldman. Theoretical considerations for THz SIS mixers. *International Journal of Infrared and Millimeter Waves*, 8:1287–1292, 1987. doi: 10.1007/BF01011080.

[4] G. P. Williams. Filling the THz gap-high power sources and applications. *Reports on Progress in Physics*, 69(2):301, 2006.

[5] W. Zhang, P. Khosropanah, J. R. Gao, T. Bansal, T. M. Klapwijk, W. Miao, and S. C. Shi. Noise temperature and beam pattern of an NbN hot electron bolometer mixer at 5.25 THz. *Journal of Applied Physics*, 108(9):093102, 2010. doi: 10.1063/1.3503279.

[6] H.-W. Hubers. Terahertz Heterodyne Receivers. *Selected Topics in Quantum Electronics, IEEE Journal of*, 14(2):378 –391, march-april 2008. doi: 10.1109/JSTQE.2007.913964.

[7] S. P. Langley. The Actinic Balance. *The American Journal of Science*, 21:187–197, March 1881.

[8] S. P. T. The Bolometer. *Nature*, 25:14–16, November 1881. doi: 10.1038/025014a0.

[9] P. L. Richards. Bolometers for infrared and millimeter waves. *Journal of applied physics*, 76(1):1–24, 1994. doi: 10.1063/1.357128.

[10] W. M. H. Haynes, editor. *2012 - 2013*. CRC handbook of chemistry and physics ; 93. CRC Press, Boca Raton, Fla., 93. ed., 2012/2013 edition, 2012. Previous ed.: 2010. - Includes bibliographical references and index.

[11] A. Subrahmanyam, Y. B. K. Reddy, and C. L. Nagendra. Nano-vanadium oxide thin films in mixed phase for microbolometer applications. *Journal of Physics D: Applied Physics*, 41(19):195108, 2008. doi: 10.1088/0022-3727/41/19/195108.

[12] A. Torres, A. Kosarev, M. G. Cruz, and R. Ambrosio. Uncooled micro-bolometer based on amorphous germanium film. *Journal of Non-Crystalline Solids*, 329(1-3):179 – 183, 2003. doi: 10.1016/j.jnoncrysol.2003.08.037. 7th Int. Workshop on Non-Crystalline Solids.

[13] J. J. Bock, D. Chen, P. D. Mauskopf, and A. E. Lange. A novel bolometer for infrared and millimeter-wave astrophysics. *Space Science Reviews*, 74:229–235, 1995. doi: 10.1007/BF00751274.

[14] C. Kittel. *Einführung in die Festkörperphysik*. Oldenbourg Verl., 2006.

[15] D. H. Andrews, J. W. F. Brucksch, W. T. Ziegler, and E. R. Blanchard. Attenuated Superconductors I. For Measuring Infra-Red Radiation. *Review of Scientific Instruments*, 13(7):281–292, 1942. doi: 10.1063/1.1770037.

[16] S.-F. Lee, J. M. Gildemeister, W. Holmes, A. T. Lee, and P. L. Richards. Voltage-Biased Superconducting Transition-Edge Bolometer with Strong Electrothermal Feedback Operated at 370 mK. *Appl. Opt.*, 37(16):3391–3397, Jun 1998. doi: 10.1364/AO.37.003391.

[17] K. D. Irwin. An application of electrothermal feedback for high resolution cryogenic particle detection. *Applied Physics Letters*, 66(15):1998–2000, 1995. doi: 10.1063/1.113674.

[18] S. Anders, T. May, V. Zakosarenko, M. Starkloff, G. Zieger, and H.-G. Meyer. Structured SiN membranes as platform for cryogenic bolometers. *Microelectronic Engineering*, 86(4 - 6):913 – 915, 2009. doi: 10.1016/j.mee.2008.10.021.

[19] Y. Al'ber, A. Andronov, V. Valov, V. Kozlov, and I. Ryazantseva. Hot electron population inversion and cyclotron resonance negative differential conductivity in semiconductors. *Solid State Communications*, 19(10):955 – 959, 1976. doi: 10.1016/0038-1098(76)90629-3.

[20] F. C. Wellstood, C. Urbina, and J. Clarke. Hot-electron effects in metals. *Phys. Rev. B*, 49:5942–5955, Mar 1994. doi: 10.1103/PhysRevB.49.5942.

[21] N. Perrin and C. Vanneste. Response of superconducting films to a periodic optical irradiation. *Phys. Rev. B*, 28:5150–5159, Nov 1983. doi: 10.1103/PhysRevB.28.5150.

[22] A. D. Semenov, G. N. Gol'tsman, and R. Sobolewski. Hot-electron effect in superconductors and its applications for radiation sensors. *Superconductor Science and Technology*, 15(4):R1, 2002. doi: 10.1088/0953-2048/15/4/201.

[23] Y. Gousev, A. Semenov, G. Gol'tsman, A. Sergeev, and E. Gershenzon. Electron-phonon interaction in disordered NbN films. *Physica B: Condensed Matter*, 194-196, Part 1(0):1355 – 1356, 1994. doi: 10.1016/0921-4526(94)91007-3.

[24] G. Bergmann. Weak localization in thin films: a time-of-flight experiment with conduction electrons. *Physics Reports*, 107(1):1 – 58, 1984. doi: 10.1016/0370-1573(84)90103-0.

[25] Y. Mushiake. Self-complementary antennas. *Antennas and Propagation Magazine, IEEE*, 34(6):23 –29, dec. 1992. doi: 10.1109/74.180638.

[26] D. E. Prober. Superconducting terahertz mixer using a transition-edge microbolometer. *Applied Physics Letters*, 62(17):2119–2121, 1993. doi: 10.1063/1.109445.

[27] L. Jiang, W. Zhang, N. Li, Z. Lin, Q. Yao, W. Miao, S. Shi, S. Svechnikov, Y. Vakhtomin, S. Antipov, B. Voronov, and G. Goltsman. Characterization of the Performance of a Quasi-Optical NbN Superconducting HEB Mixer. *Applied Superconductivity, IEEE Transactions on*, 17(2):395 –398, june 2007. doi: 10.1109/TASC.2007.897875.

[28] B. J. Dalrymple, S. A. Wolf, A. C. Ehrlich, and D. J. Gillespie. Inelastic electron lifetime in niobium films. *Phys. Rev. B*, 33:7514–7519, Jun 1986. doi: 10.1103/PhysRevB.33.7514.

[29] B. S. Karasik, K. S. Il'in, E. V. Pechen, and S. I. Krasnosvobodtsev. Diffusion cooling mechanism in a hot-electron NbC microbolometer mixer. *Applied Physics Letters*, 68(16):2285–2287, 1996. doi: 10.1063/1.115886.

[30] P. J. Burke, R. J. Schoelkopf, D. E. Prober, A. Skalare, W. R. McGrath, B. Bumble, and H. G. LeDuc. Length scaling of bandwidth and noise in hot-electron superconducting mixers. *Applied Physics Letters*, 68(23):3344–3346, 1996. doi: 10.1063/1.116052.

[31] A. Semenov, B. Günther, U. Böttger, H.-W. Hübers, H. Bartolf, A. Engel, A. Schilling, K. Ilin, M. Siegel, R. Schneider, D. Gerthsen, and N. A. Gippius. Optical and transport properties of ultrathin NbN films and nanostructures. *Phys. Rev. B*, 80:054510, Aug 2009. doi: 10.1103/PhysRevB.80.054510.

[32] K. S. Il'in, M. Lindgren, M. Currie, A. D. Semenov, G. N. Gol'tsman, R. Sobolewski, S. I. Cherednichenko, and E. M. Gershenzon. Picosecond hot-electron energy relaxation in NbN superconducting photodetectors. *Applied Physics Letters*, 76(19):2752–2754, 2000. doi: 10.1063/1.126480.

[33] S. B. Kaplan. Acoustic matching of superconducting films to substrates. *Journal of Low Temperature Physics*, 37:343–365, 1979. doi: 10.1007/BF00119193.

[34] D. Wilms Fleot, E. Miedema, J. Baselmans, T. Klapwijk, and J. Gao. Resistive states of superconducting hot-electron bolometer mixers: charge-imbalance vs. hotspot. *Applied Superconductivity, IEEE Transactions on*, 9(2):3749 –3752, jun 1999. doi: 10.1109/77.783843.

[35] W. J. Skocpol, M. R. Beasley, and M. Tinkham. Self-heating hotspots in superconducting thin-film microbridges. *Journal of Applied Physics*, 45(9):4054–4066, 1974. doi: 10.1063/1.1663912.

[36] P. Khosropanah, H. Merkel, S. Yngvesson, A. Adam, S. Cherednichenko, and E. Kollberg. A Distributed Device Model for Phonon-Cooled HEB Mixers Pre-

dicting IV Characteristics, Gain, Noise and IF Bandwidth. In *Eleventh International Symposium on Space Terahertz Technology*, page 474. May 2000.

[37] A. D. Semenov, R. S. Nebosis, Y. P. Gousev, M. A. Heusinger, and K. F. Renk. Analysis of the nonequilibrium photoresponse of superconducting films to pulsed radiation by use of a two-temperature model. *Phys. Rev. B*, 52:581–590, Jul 1995. doi: 10.1103/PhysRevB.52.581.

[38] E. Gerecht, C. Musante, Y. Zhuang, K. Yngvesson, G. Gol'tsman, B. Voronov, and E. Gershenzon. NbN hot electron bolometric mixers-a new technology for low-noise THz receivers. *Microwave Theory and Techniques, IEEE Transactions on*, 47(12):2519–2527, 1999. doi: 10.1109/22.809001.

[39] A. Scheuring. Entwicklung ultra-breitbandiger Antennenstrukturen für supraleitende und normalleitende Terahertz Strahlungsdetektoren. Ph.D. thesis, Karlsruhe Institut für Technologie, 2013.

[40] C. A. Balanis. *Antenna Theory: Analysis and Design*. Wiley-Interscience, 2005.

[41] F. J. Low and A. R. Hoffman. The Detectivity of Cryogenic Bolometers. *Appl. Opt.*, 2(6):649–650, Jun 1963. doi: 10.1364/AO.2.000649.

[42] J. C. Mather. Bolometer noise: nonequilibrium theory. *Applied Opttics*, 21:1125–1129, 1982. doi: 10.1364/AO.21.001125.

[43] A. Kerr. Suggestions for revised definitions of noise quantities, including quantum effects. *Microwave Theory and Techniques, IEEE Transactions on*, 47(3):325 –329, mar 1999. doi: 10.1109/22.750234.

[44] H. B. Callen and T. A. Welton. Irreversibility and Generalized Noise. *Phys. Rev.*, 83:34–40, Jul 1951. doi: 10.1103/PhysRev.83.34.

[45] B. Karasik, W. McGrath, and R. Wyss. Optimal choice of material for HEB superconducting mixers. *Applied Superconductivity, IEEE Transactions on*, 9(2):4213 –4216, jun 1999. doi: 10.1109/77.783954.

[46] A. D. Semenov, K. Ilin, M. Siegel, A. Smirnov, S. Pavlov, H. Richter, and H.-W. Hübers. Evidence of non-bolometric mixing in the bandwidth of a hot-electron

bolometer. *Superconductor Science and Technology*, 19(10):1051, 2006. doi: 10.1088/0953-2048/19/10/011.

[47] S. Cherednichenko, V. Drakinskiy, J. Baubert, J.-M. Krieg, B. Voronov, G. Gol'tsman, and V. Desmaris. Gain bandwidth of NbN hot-electron bolometer terahertz mixers on 1.5 mu m Si[sub 3]N[sub 4]/SiO[sub 2] membranes. *Journal of Applied Physics*, 101(12):124508, 2007. doi: 10.1063/1.2749302.

[48] K. Ilin, R. Schneider, D. Gerthsen, A. Engel, H. Bartolf, A. Schilling, A. Semenov, H.-W. Huebers, B. Freitag, and M. Siegel. Ultra-thin NbN films on Si: crystalline and superconducting properties. *Journal of Physics: Conference Series*, 97(1):012045, 2008. doi: 10.1088/1742-6596/97/1/012045.

[49] D. Doenitz, R. Kleiner, D. Koelle, T. Scherer, and K. F. Schuster. Imaging of thermal domains in ultrathin NbN films for hot electron bolometers. *Applied Physics Letters*, 90(25):252512, 2007. doi: 10.1063/1.2751109.

[50] J. Baselmans, M. Hajenius, J. Gao, A. Baryshev, J. Kooi, T. Klapwijk, B. Voronov, P. de Korte, and G. Gol'tsman. NbN hot electron bolometer mixers: sensitivity, LO power, direct detection and stability. *Applied Superconductivity, IEEE Transactions on*, 15(2):484–489, June 2005. doi: 10.1109/TASC.2005.849884.

[51] D. Meledin, C.-Y. Tong, R. Blundell, N. Kaurova, K. Smirnov, B. Voronov, and G. Goltsman. Study of the IF bandwidth of NbN HEB mixers based on crystalline quartz substrate with an MgO buffer layer. *Applied Superconductivity, IEEE Transactions on*, 13(2):164 – 167, june 2003. doi: 10.1109/TASC.2003.813671.

[52] W. Zhang, J.-R. Gao, M. Hajenius, W. Miao, P. Khosropanah, T. Klapwijk, and S.-C. Shi. Twin-Slot Antenna Coupled NbN Hot Electron Bolometer Mixer at 2.5 THz. *Terahertz Science and Technology, IEEE Transactions on*, 1(2):378 –382, nov. 2011. doi: 10.1109/TTHZ.2011.2145330.

[53] R. Schneider, B. Freitag, D. Gerthsen, K. S. Ilin, and M. Siegel. Structural, microchemical and superconducting properties of ultrathin NbN films

120

on silicon. *Crystal Research and Technology*, 44(10):1115–1121, 2009. doi: 10.1002/crat.200900462.

[54] L. N. Cooper. Superconductivity in the Neighborhood of Metallic Contacts. *Phys. Rev. Lett.*, 6(12):689–690, Jun 1961. doi: 10.1103/PhysRevLett.6.689.

[55] S. Wolf, J. Kennedy, and M. Nisenoff. Properties of superconducting rf sputtered ultra thin films of Nb. *Applied Superconductivity, IEEE Transactions on*, 13(21):145–147, January 1976. doi: 10.1116/1.568809.

[56] K. Fuchs. The conductivity of thin metallic films according to the electron theory of metals. *Mathematical Proceedings of the Cambridge Philosophical Society*, 34(01):100–108, 1938. doi: 10.1017/S0305004100019952.

[57] A. D. Semenov, H.-W. Hubers, J. Schubert, G. N. Gol'tsman, A. I. Elantiev, B. M. Voronov, and E. M. Gershenzon. Design and performance of the lattice-cooled hot-electron terahertz mixer. *Journal of Applied Physics*, 88(11):6758–6767, 2000. doi: 10.1063/1.1323531.

[58] J. J. A. Baselmans, M. Hajenius, J. R. Gao, T. M. Klapwijk, P. A. J. de Korte, B. Voronov, and G. Gol'tsman. Doubling of sensitivity and bandwidth in phonon cooled hot electron bolometer mixers. *Applied Physics Letters*, 84(11):1958–1960, 2004. doi: 10.1063/1.1667012.

[59] Y. V. Fominov and M. V. Feigel'man. Superconductive properties of thin dirty superconductor-normal-metal bilayers. *Phys. Rev. B*, 63(9):094518, Feb 2001. doi: 10.1103/PhysRevB.63.094518.

[60] K. Il'in, A. Stockhausen, A. Scheuring, M. Siegel, A. Semenov, H. Richter, and H.-W. Huebers. Technology and Performance of THz Hot-Electron Bolometer Mixers. *Applied Superconductivity, IEEE Transactions on*, 19(3):269 –273, june 2009. doi: 10.1109/TASC.2009.2018266.

[61] J. Schubert, A. Semenov, G. Gol'tsman, H. W. Hübers, G. Schwaab, B. Voronov, and E. Gershenzon. Noise Temperature and Sensitivity of a NbN Hot-Electron Mixer at Frequencies from0.7 THz to 5.2 THz. In *Proceedings on the tenth International Symposium on Space Terahertz Technology*, pages 190–199. 1999.

[62] P. Khosropanah, J. R. Gao, W. M. Laauwen, M. Hajenius, and T. M. Klapwijk. Low noise NbN hot electron bolometer mixer at 4.3 THz. *Applied Physics Letters*, 91(22):221111, 2007.

[63] P. Khosropanah, W. Zhang, J. N. Hovenier, J. R. Gao, T. M. Klapwijk, M. I. Amanti, G. Scalari, and J. Faist. 3.4 THz heterodyne receiver using a hot electron bolometer and a distributed feedback quantum cascade laser. *Journal of Applied Physics*, 104(11):113106, 2008. doi: 10.1063/1.3032354.

[64] E. Gerecht, C. Musante, Y. Zhuang, M. Ji, K. Yngvesson, T. Goyette, and J. Waldman. NbN hot electron bolometric mixer with intrinsic receiver noise temperature of less than five times the quantum noise limit. In *Microwave Symposium Digest. 2000 IEEE MTT-S International*, volume 2, pages 1007 –1010 vol.2. 2000. doi: 10.1109/MWSYM.2000.863527.

[65] E. Gerecht, C. F. Musante, H. Jian, K. S. Yngvesson, J. Dickinson, J. Waldman, G. N. Gol'tsman, P. A. Yagoubov, B. M. Voronov, and E. M. Gershenzon. Measured Results for NbN Phonon-Cooled Hot Electron Bolometric Mixers at 0.6-0.75 THz, 1.56 THz, and 2.5 THz. In *Proceedings on the nineth International Symposium on Space Terahertz Technology*, pages 105–114. 1998.

[66] E. Gerecht, C. F. Musante, H. Jian, K. S. Yngvesson, J. Dickinson, J. Waldman, G. N. Gol'tsman, P. A. Yagoubov, B. M. Voronov, and E. M. Gershenzon. Development of Focal Plane Arrays Utilizing NbN Hot Electron Bolometric Mixers for the THz Regime. In *Proceedings on the 11th International Symposium on Space Terahertz Technology*, pages 209–218. 2000.

[67] M. Kroug, S. Cherednichenko, H. Merkel, E. Kollberg, B. Voronov, G. Gol'tsman, H. Huebers, and H. Richter. NbN hot electron bolometric mixers for terahertz receivers. *Applied Superconductivity, IEEE Transactions on*, 11(1):962 –965, mar 2001. doi: 10.1109/77.919508.

[68] S. Miki, Y. Uzawa, A. Kawakami, and Z. Wang. IF bandwidth and noise temperature measurements of NbN HEB mixers on MgO substrates. *Applied Superconductivity, IEEE Transactions on*, 11(1):175 –178, mar 2001. doi: 10.1109/77.919313.

[69] H. Kataoka, A. Kawakami, Y. Uzawa, Z. Wang, and N. Kaya. Fabrication of NbN-HEB mixers with fluoride radical etching Process. *Applied Superconductivity, IEEE Transactions on*, 15(2):469 – 471, june 2005. doi: 10.1109/TASC.2005.849877.

[70] J. Chen, M. Liang, L. Kang, B. Jin, W. Xu, P. Wu, W. Zhang, L. Jiang, N. Li, and S. Shi. Low Noise Receivers at 1.6 THz and 2.5 THz Based on Niobium Nitride Hot Electron Bolometer Mixers. *Applied Superconductivity, IEEE Transactions on*, 19(3):278 –281, june 2009. doi: 10.1109/TASC.2009.2017910.

[71] Y. Delorme, R. Lefèvre, W. Miao, A. Féret, W. Zhang, T. V. andF. Dauplay, L. Pelay, J. Spatazza, M. B. Trung, J.-M. Krieg, Y.Jin, P. Khosropanah, J. R. Gao, and S. C. Shi. A Quasi-Optical NbN Mixer with 800K DSB Noise Temperature at 2.5 THz. In *Proceedings on the 22nd International Symposium on Space Terahertz Technology*, pages 135–138. 2011.

[72] I. Tretyakov, S. Ryabchun, M. Finkel, A. Maslennikova, N. Kaurova, A. Lobastova, B. Voronov, and G. Gol'tsman. Low noise and wide bandwidth of NbN hot-electron bolometer mixers. *Applied Physics Letters*, 98(3):033507, 2011. doi: 10.1063/1.3544050.

[73] B. S. Karasik, M. C. Gaidis, W. R. McGrath, B. Bumble, and H. G. LeDuc. Low noise in a diffusion-cooled hot-electron mixer at 2.5 THz. *Applied Physics Letters*, 71(11):1567–1569, 1997. doi: 10.1063/1.119967.

[74] M. J. Myers, W. Holzapfel, A. T. Lee, R. O'Brient, P. L. Richards, H. T. Tran, P. Ade, G. Engargiola, A. Smith, and H. Spieler. An antenna-coupled bolometer with an integrated microstrip bandpass filter. *Applied Physics Letters*, 86(11):114103, 2005. doi: 10.1063/1.1879115.

[75] S. Yamasaki and T. Aomine. Self-Heating Effects in Long Superconducting Thin Films over a Wide Temperature Range. *Japanese Journal of Applied Physics*, 18(3):667–670, 1979. doi: 10.1143/JJAP.18.667.

[76] A. Stockhausen, K. Il'in, M. Siegel, U. Södervall, P. Jedrasik, A. Semenov, and H.-W. Hübers. Adjustment of self-heating in long superconducting thin film NbN

microbridges. *Superconductor Science and Technology*, 25(3):035012, 2012. doi: 10.1088/0953-2048/25/3/035012.

[77] J. Callaway. Model for Lattice Thermal Conductivity at Low Temperatures. *Phys. Rev.*, 113(4):1046–1051, Feb 1959. doi: 10.1103/PhysRev.113.1046.

[78] L. Weber and E. Gmelin. Transport properties of silicon. *Applied Physics A: Materials Science & Processing*, 53:136–140, 1991. doi: 10.1007/BF00323873.

[79] Y. F. Zhu, J. S. Lian, and Q. Jiang. Re-examination of Casimir limit for phonon traveling in semiconductor nanostructures. *Applied Physics Letters*, 92(11):113101, 2008. doi: 10.1063/1.2898516.

[80] F. Laermer and A. Schilp. Verfahren zum anisotropen Ätzen von Silicium. Patent, May 1994.

[81] M. W. Johnson, A. M. Herr, and A. M. Kadin. Bolometric and nonbolometric infrared photoresponses in ultrathin superconducting NbN films. *Journal of Applied Physics*, 79(9):7069–7074, 1996. doi: 10.1063/1.361426.

[82] A. Semenov, H.-W. Hubers, K. Il'in, M. Siegel, V. Judin, and A.-S. Muller. Monitoring coherent THz-synchrotron radiation with superconducting NbN hot-electron detector. In *Infrared, Millimeter, and Terahertz Waves, 2009. IRMMW-THz 2009. 34th International Conference on*, pages 1 –2. sept. 2009. doi: 10.1109/ICIMW.2009.5324688.

[83] B. S. Karasik and A. I. Elantiev. Noise temperature limit of a superconducting hot-electron bolometer mixer. *Applied Physics Letters*, 68(6):853–855, 1996. doi: 10.1063/1.116555.

[84] R. Yuan, M. Wei, Y. Qi-Jun, Z. Wen, and S. Sheng-Cai. Terahertz Direct Detection Characteristics of a Superconducting NbN Bolometer. *Chinese Physics Letters*, 28(1):010702, 2011.

[85] F. R. Elder, A. M. Gurewitsch, R. V. Langmuir, and H. C. Pollock. Radiation from Electrons in a Synchrotron. *Phys. Rev.*, 71:829–830, Jun 1947. doi: 10.1103/PhysRev.71.829.5.

[86] P. Probst, A. Scheuring, M. Hofherr, D. Rall, S. Wünsch, K. Il'in, M. Siegel, A. Semenov, A. Pohl, H.-W. Hübers, V. Judin, A.-S. Müller, A. Hoehl, R. Müller, and G. Ulm. YBa$_2$Cu$_3$O$_{7-\delta}$ quasioptical detectors for fast time-domain analysis of terahertz synchrotron radiation. *Applied Physics Letters*, 98(4):043504, 2011. doi: 10.1063/1.3546173.

[87] A.-S. Müller, Y.-L. Mathis, I. Birkel, B. Gasharova, C. Hirschmugl, E. Huttel, D. Moss, R. Rossmanith, and P. Wesolowski. Far Infrared Coherent Synchrotron Edge Radiation at ANKA. *Synchrotron Radiation News*, 19(3):18–24, 2006. doi: 10.1080/08940880600755202.

[88] H.-W. Hübers, A. Semenov, K. Holldack, U. Schade, G. Wüstefeld, and G. Gol'tsman. Time domain analysis of coherent terahertz synchrotron radiation. *Applied Physics Letters*, 87(18):184103, 2005. doi: 10.1063/1.2120896.

[89] S. Cherednichenko, P. Khosropanah, E. Kollberg, M. Kroug, and H. Merkel. Terahertz superconducting hot-electron bolometer mixers. *Physica C: Superconductivity*, 372-376, Part 1(0):407 – 415, 2002. doi: 10.1016/S0921-4534(02)00711-6.

[90] J. Faist, F. Capasso, D. L. Sivco, C. Sirtori, A. L. Hutchinson, and A. Y. Cho. Quantum Cascade Laser. *Science*, 264(5158):553–556, 1994. doi: 10.1126/science.264.5158.553.

[91] S. P. Khanna, M. Salih, P. Dean, A. G. Davies, and E. H. Linfield. Electrically tunable terahertz quantum-cascade laser with a heterogeneous active region. *Applied Physics Letters*, 95(18):181101, 2009. doi: 10.1063/1.3253714.

[92] T. Crowe, W. Bishop, D. Porterfield, J. Hesler, and I. Weikle, R.M. Opening the terahertz window with integrated diode circuits. *Solid-State Circuits, IEEE Journal of*, 40(10):2104 – 2110, oct. 2005. doi: 10.1109/JSSC.2005.854599.

[93] A. Scheuring, P. Dean, A. Valavanis, A. Stockhausen, P. Thoma, M. Salih, S. Khanna, S. Chowdhury, J. Cooper, A. Grier, S. Wuensch, K. Il'in, E. Linfield, A. Davies, and M. Siegel. Transient Analysis of THz-QCL Pulses Using

NbN and YBCO Superconducting Detectors. *Terahertz Science and Technology, IEEE Transactions on*, 3(2):172–179, 2013. doi: 10.1109/TTHZ.2012.2228368.

[94] S. Cibella, M. Beck, P. Carelli, M. Castellano, F. Chiarello, J. Faist, R. Leoni, M. Ortolani, L. Sabbatini, G. Scalari, G. Torrioli, and D. Turcinkova. Operation of a Wideband Terahertz Superconducting Bolometer Responding to Quantum Cascade Laser Pulses. *Journal of Low Temperature Physics*, pages 1–6, 2011. doi: 10.1007/s10909-012-0588-5.

Lebenslauf

13. Juli 1981 Geboren in Heilbronn

1987 - 1991 Besuch der Gerhart-Hauptmann Schule, Heilbronn

1991 - 2000 Besuch des Theodor-Heuss-Gymnasium, Heilbronn

2000 Abitur

2000 - 2001 Zivildienst

2001 - 2007 Studium der Physik an der Universität Karlsruhe (TH)

2007 Diplom in Physik

2007 - 2012 Wissenschaftlicher Mitarbeiter
am Institut für Mikro- und Nanoelektronische Systeme,
Karlsruhe Institut für Technologie (KIT)

Karlsruher Schriftenreihe zur Supraleitung
(ISSN 1869-1765)

Herausgeber: Prof. Dr.-Ing. M. Noe, Prof. Dr. rer. nat. M. Siegel

Die Bände sind unter www.ksp.kit.edu als PDF frei verfügbar oder als Druckausgabe bestellbar.

Band 001
Christian Schacherer
Theoretische und experimentelle Untersuchungen zur Entwicklung supraleitender resistiver Strombegrenzer. 2009
ISBN 978-3-86644-412-6

Band 002
Alexander Winkler
Transient behaviour of ITER poloidal field coils. 2011
ISBN 978-3-86644-595-6

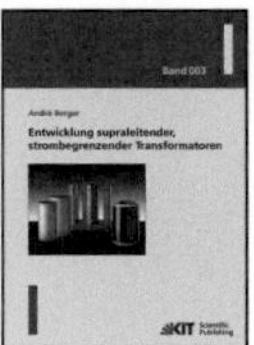

Band 003
André Berger
Entwicklung supraleitender, strombegrenzender Transformatoren. 2011
ISBN 978-3-86644-637-3

Band 004
Christoph Kaiser
High quality Nb/Al-AlOx/Nb Josephson junctions. Technological development and macroscopic quantum experiments. 2011
ISBN 978-3-86644-651-9

Band 005
Gerd Hammer
Untersuchung der Eigenschaften von planaren Mikrowellenresonatoren für Kinetic-Inductance Detektoren bei 4,2 K. 2011
ISBN 978-3-86644-715-8

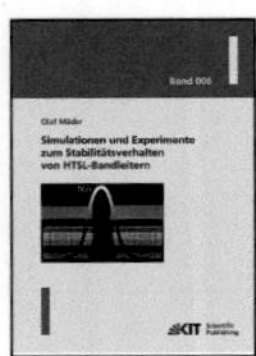

Band 006
Olaf Mäder
**Simulationen und Experimente zum Stabilitätsverhalten
von HTSL-Bandleitern.** 2012
ISBN 978-3-86644-868-1

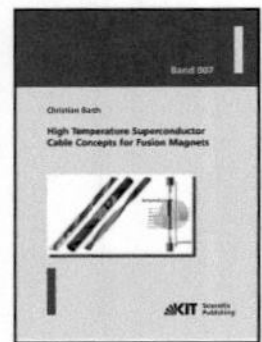

Band 007
Christian Barth
**High Temperature Superconductor Cable Concepts for
Fusion Magnets.** 2013
ISBN 978-3-7315-0065-0

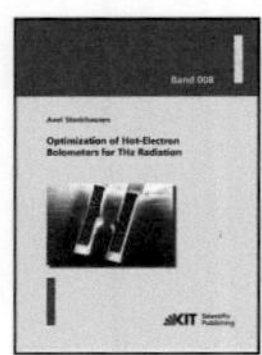

Band 008
Axel Stockhausen
Optimization of Hot-Electron Bolometers for THz Radiation. 2013
ISBN 978-3-7315-0066-7